Maxime Kenfack

The seal of the atom

Maxime Kenfack

The seal of the atom

The Sovereignty Manual

ScienciaScripts

Imprint

Cover image: www.ingimage.com

This book is a translation from the original published under ISBN 978-620-9-67357-3.

Publisher:
Sciencia Scripts
is a trademark of
Dodo Books Indian Ocean Ltd. and OmniScriptum S.R.L publishing group

120 High Road, East Finchley, London, N2 9ED, United Kingdom
Str. Armeneasca 28/1, office 1, Chisinau MD-2012, Republic of Moldova, Europe
Managing Directors: Ieva Konstantinova, Victoria Ursu
info@omniscriptum.com

Printed at: see last page
ISBN: 978-620-9-71942-4

Table of Contents

PUBLICATION ARGUMENT: THE SEAL OF THE ATOM

I. Book Specifications

- **Title:** THE SEAL OF THE ATOM: The Manual of Sovereignty.
- **Author:** Maxime KENFACK.
- **Target Publisher:** European University Presses (EUE).
- **Nature:** 13th official manuscript in the author's catalog.
- **Genre:** Geopolitical Essay / Treatise on Technological and Nuclear Prospective.
- **Structure:** 40 Books (Theoretical Chapters) linking atomic power and algorithmic intelligence.

II. Analytical Summary

"THE SEAL OF THE ATOM" offers a high-level reflection on the resilience of modern nations. At the intersection of nuclear physics and data science, Maxime KENFACK theorizes **the World-Object** as a structure of integral sovereignty. The author demonstrates that France, to maintain its standing, must forge an alliance between its historical mastery of **the Atom** (energy and deterrence) and its conquest of **Silica** (artificial intelligence).

Drawing on a rich life experience—from the traditional wisdom of **Foto** to fifteen years of civic engagement with the **French Red Cross**—the author proposes a model of "Sovereignty of Peace" in which technology becomes the armed wing of humanism.

III. Academic Value and Key Strengths

- **Cross-Disciplinary Expertise:** The book bridges often compartmentalized fields: nuclear strategy, quantum cryptography, and republican ethics.

- **Originality of Thought:** The introduction of the concept of **"Restoration of Souls"** into a technological treatise offers a novel sociological perspective, essential for European University Presses.
- **Depth of the Catalog:** As **the 13th volume**, this text crowns a cycle of reflection begun with **The Seal of the Pen* and **The Digital Curtain**, confirming the author's stature as an independent researcher and prolific essayist.
- **Geopolitical Relevance:** This handbook directly addresses the challenges of European strategic autonomy in the face of global technological blocs.

IV. Target Audience (Academic)

- Researchers and doctoral students in Political Science and International Relations.

- Engineers and experts in the energy and cybersecurity sectors.
- Senior civil servants and judges of the Administrative Court (as evidence of intellectual contribution to French heritage).
- University libraries and strategic research centers.

GENERAL INTRODUCTION: THE MANIFESTO OF THE WORLD-OBJECT

I. The Genesis of a Vision: From Village to Republic

The idea for this manual did not originate in high-tech laboratories, but on the red dirt roads of Cameroon, inspired by the wisdom of **Martin** and **Pauline**. In Foto, sovereignty was a matter of community survival, respect for elders, and mastery of the land. Fifteen years after my arrival in France, this vision encountered scientific excellence and republican rigor.

The conclusion is clear: in a turbulent world, a nation that does not possess its own energy (**the Atom**) and its own intelligence (**Silicon**) is doomed to fade into obscurity.

II. The Equation of Sovereignty: $P = (E + I) \times L$

In this book, I propose a new law of national power, where Power (P) is the product of Energy (E) and Intelligence (I), multiplied by the Legitimacy of the Law (L).

- **The Atom** provides raw physical force.
- **Silica** provides the capacity for calculation and foresight.
- **The Seal of the Pen** ensures that this power remains in the service of humanity, peace, and justice.

III. Why 40 Books? The Structure of a Renaissance

This treatise is divided into forty chapters, for sovereignty is a complete edifice. One cannot protect cybersecurity (Book XXX) without mastering the forest (Book XXIV) or the sea (Book XXVII). Each book is a stone in **the World-Object**, that global system where technology and humanism are one.

My volunteer work with **Secours Catholique** and the **Red Cross** has taught me that precariousness often stems from a loss of control over basic resources. This book is my intellectual response to the fragility of the world.

IV. The Author's Mission: To Restore Souls Through Reason

As a writer, my craft and I are one. My pen is a tool of peace. By publishing this 10th manuscript, I wish to offer France—my adopted homeland and the home of my heart—a navigation manual for the coming century. It is a tribute to French culture, its language, and its universal capacity to illuminate the shadows.

Scientific Statement of Intent

1. **Book Project:** *The Atom and Silica: Epistemology of Nuclear Sovereignty in the*

Age of ***Hyper-Connectivity***

I. Central Issue

At the dawn of the 21st century, the doctrine of nuclear deterrence, a pillar of global stability since 1945, is confronting a disruptive technological paradigm: the omnipresence of Silica. While atomic power has historically rested on mastery of the atom, its implementation, integrity, and credibility now depend on the digital realm. This book project aims to analyze the critical correlation between **energy/military sovereignty** and **digital sovereignty**, positing that the vulnerability of the information vector directly threatens the balance of terror.

II. Research Objectives

The study aims to deconstruct the concept of "Regilian power" along three major academic axes:

1. **Analysis of cybernetic dependency:** How is the architecture of command and control (C2) systems structurally linked to the integrity of semiconductors?
2. **The Paradox of Digital Deterrence:** In an era where cyber sabotage (Stuxnet, etc.) can neutralize a strike force without a declaration of war, how can we redefine the sovereignty of a nuclear state?
3. **Modeling the Resilience Protocol:**
   Proposing a theory of "Integral Sovereignty" in which the atom and computer code form a block of invulnerability.

III. Methodology and Theoretical Contribution

This work builds upon the theory of **the World-Object**. It employs an interdisciplinary approach that combines classical geopolitics, social physics, and complex systems engineering. The original contribution lies in the introduction of the **"Theory of the Atom's Seal,"** a mathematical and philosophical model of the technical trust necessary for maintaining world peace.

IV. Prospective Conclusion

The book will demonstrate that a nation's sovereignty is no longer measured solely by its kinetic arsenal, but by its ability to sanctify its digital space. It will conclude with the necessity of a new global ethic of technological responsibility.

BOOK I:

THE ONTOLOGY OF THE " " ATOMIC POWER AND THE METAMORPHOSIS OF " " SOVEREIGNTY

I. The Atom as an Epistemological Break with Territory

Sovereignty, as inherited from the Treaties of Westphalia, rested on the linearity of borders and the tangibility of kinetic force. I argue that the advent of the atom in 1945 constitutes not merely an evolution in weaponry, but a transformation of the very ontology of the state. For the first time in human history, power is no longer measured by the capacity for conquest, but by the capacity to annihilate the adversary's space-time.

The nuclear state becomes a **"Sanctuary-State**. " Here I analyze how the border loses its nature as a customs line or barbed wire to become an aura of invulnerability. In this first volume, I examine how the atomic object has redefined the social contract: the citizen no longer entrusts his security to an organic force, but to a technological abstraction capable of triggering the apocalypse.

II. The Paradox of Deterrence: Between Silence and Speech

Atomic **power** is the only force that exists only as long as it is not used. I define it as a **"power of the word**." Here I explore the philosophical dimension of deterrence: how does the word of a head of state become the trigger for an invisible force?

I put this reality into perspective with my earlier work on **"The World-Object**." If the atom is the ultimate object, I demonstrate that it is also the most fragile object, for it rests entirely on the credibility of communication. It is here that I establish the critical link with my research on the digital: if the message is corrupted, the power is null.

III. The Geopolitics of the Sacred: From Foto to the Security Council

I perceive a quasi-mystical dimension in the possession of nuclear fire. I develop a comparative analysis between the ancestral management of the sacred—as transmitted to me by the wisdom of my village of **Foto**—and the modern management of the ultimate weapon.

The forbidden, the taboo, responsibility toward future generations: I postulate that these concepts are not merely ethical, but constitute the very cement of global stability. A state that possesses the atom bears a responsibility that transcends its own national destiny to touch upon the universal. It is this dimension of **the "Keeper of the Fire"** that I wish to theorize here.

IV. The Genesis of Integral Sovereignty: My Interpretation of the French Model (1958–2026)

1. The Epistemology of Independence: My Analysis of the Gaullist Turn

I assert that French sovereignty cannot be understood without a rigorous deconstruction of the concept of the "National Sanctuary." In my work, I postulate that a nation's survival is an inalienable prerogative. Here, I observe the transition from a traditional power to a **sovereign technological power**. Through the study of the CEA and the CNES, I demonstrate that scientific research is not an accessory of the State, but the very engine of its intellectual renaissance.

2. The "Weak Against the Strong" Doctrine: My Model of Deterrence

In this demonstration, I develop the mathematical logic underlying a middle power's ability to paralyze a superpower. Here I introduce the concept of **the Unacceptable Damage Threshold**. I argue that nuclear force constitutes the first true physical "World-Object": its mere existence, as I define it, reconfigures the behavioral space of all global actors.

3. The Digital Transformation of Strike Capability: The Contribution of My Protocols

I now come to the crux of my argument: the intrinsic vulnerability of the delivery system. I analyze the growing dependence of transmission systems on **silicon**. I firmly maintain that if electronic components escape national sovereignty, the atom itself loses its function as a sanctuary. It is here that I integrate my **Resilience Protocols** to propose absolute imperviousness for defense networks.

IV. The Dialectic of Power: From Kinetic Force to Immaterial Force

I argue that the transition to the atomic age has led to a dematerialization of war. Traditionally, sovereignty was defended through numbers and mass (kinetic force). In my model, I demonstrate that nuclear sovereignty now rests on a **"geopolitics of the mind.**"

Weapons no longer act through explosion, but through the fear of explosion. Here I analyze this semantic shift: territory is no longer protected by physical ramparts, but by a psychological and technological barrier. This is what I call the **"Frontier of the Invisible**." For a researcher, it is fascinating to observe that the more destructive the weapon, the more it becomes a pure object of language and signaling.

V. The Atom-Silica Nexus: The Vulnerability of the Sanctuary

I will now turn to the technical core of my thesis: the atom's organic dependence on **silicon**. I argue that the nuclear sanctuary is now threatened by a vulnerability that 20th-century strategists did not foresee: digital intrusion.

In my research, I identify a dangerous paradox: while the bomb is protected by tons of concrete and steel, its brain (the command and control systems) relies on micro-components of extreme fragility. I demonstrate that if the command chain is infiltrated by malicious code, the state's sovereignty collapses without a single shot being fired.

I propose here a new doctrine: **"Zero-Layer Sovereignty**." For the atom to remain a guarantee of peace, the state must imperatively control the manufacturing and integrity of its electronic components. Without digital sovereignty, nuclear sovereignty is nothing but an illusion of power.

VI. The Researcher's Responsibility: An Ethics of Vigilance

As a writer and theorist, I refuse to limit my analysis to cold technicality alone. I draw upon the ethics of the **Restoration of Souls** to question the purpose of this power. The researcher is not a neutral spectator; he is the guardian of a conscience.

I maintain that the possession of nuclear fire imposes an "**Asceticism of Power**." Drawing on the wisdom I received at **Foto**, I propose that the UN Security Council be not merely a club of military powers, but a circle of collective wisdom. I conclude this section by asserting that the atom must not be an end in itself, but the ultimate means of preserving life against the instability of the digital world.

BOOK II:

CYBERSECURITY OF CRITICAL INFRASTRUCTURE AND THE SENTINEL PROTOCOL

I. The Architecture of Vulnerability: An Autopsy of Command and Control (C2) Systems

In this section, I undertake a technical analysis of what I call the "Nervous System of Power." I postulate that a state's nuclear sovereignty no longer rests on the power of the warhead, but on the absolute integrity of the command transmission chain. Here, I analyze command and **control (C2)** networks as hybrid structures where the physical meets the digital.

I demonstrate that the introduction of **silicon** into these infrastructures has created an unprecedented attack surface. Here, I examine the vectors of compromise: electromagnetic espionage, the injection of malicious code into factory components, and the vulnerability of isolated networks (air-gaps). I argue that the nuclear sanctuary is no longer an island isolated from the world; it is, by nature, connected to a global technological value chain over which it no longer has complete control.

II. The Source Code Paradox: Sovereignty or Algorithmic Dependence

I assert that a state that uses source code it has not fully written or audited forfeits its own sovereignty. In my analysis, I examine the issue of "backdoors" and so-called "zero-day" vulnerabilities. I pose a fundamental question: can we speak of French or European deterrence if encryption or guidance algorithms depend on foreign patents?

Here I develop my theory on **"Régilian Intellectual Property**." For me, the computer code that manages the atom must be treated with the same level of secrecy and sovereignty as the formula for uranium enrichment. I propose a model for a sovereign operating system, completely disconnected from commercial standards, to guarantee the sanctity of the sanctuary.

III. The Emergence of the Sentinel Protocol: Toward Digital Immunity

Here I present my major contribution to national security: the **Sentinel Protocol**. Unlike traditional defense systems that react to an attack, I am designing a proactive protocol based on systemic resilience.

I draw on my research into **blockchain** and the decentralization of evidence to envision a tamper-proof system for validating launch orders. I explain how distributed ledger technology can be adapted to secure the transmission of a nuclear order, ensuring that no entity (whether human or malicious software) can usurp sovereign authority. The Sentinelle Protocol thus becomes the "Lock of the World-Object," ensuring that nuclear fire remains under the exclusive control of political will.

III. The Emergence of the Sentinel Protocol: Toward Systemic Digital Immunity

1. The Theory of Consensus of Fire: Beyond Human Validation

In my model, the **Sentinel Protocol** does not merely encrypt a message; it creates a multi-factor, decentralized validation ecosystem. I postulate that the nuclear command must no longer rely on a simple electrical impulse, but on a **Consensus of Sovereignty**. Here, I draw on distributed ledger technologies (Blockchain) to design a system where every communication node of the Strike Force verifies the integrity of the command in real time.

If a single line of code is altered by an external agent (malware or a zero-day intrusion), the protocol instantly detects the "object corruption" and locks down the system. This is what I call the **"Regilian Proof of Authenticity**."

2. The Resilience Algorithm: Security Through Post-Quantum Cryptography

I am integrating a layer of protection against future threats, particularly quantum computing, into the Sentinel Protocol. I maintain that current deterrence systems are obsolete in the face of tomorrow's computing power.

I am therefore developing an architecture based on asymmetric computing networks, where each command transmission is fragmented into "informational atoms" circulating on variable frequencies. To an attacker, the message is invisible, as it does not exist in its entirety anywhere until its final reconstruction by the legitimate receiver. This is the direct application of my theory of **the "Fragmented World-Object**."

3. The Sanctified Human Factor: The Human-Machine Duality

I reject the idea of total automation of nuclear fire. My protocol reintroduces the human, not as a mere executor, but as the **Ethical Key**. The Sentinel Protocol requires a "biometric signature of consciousness."

Drawing inspiration from the rites of power transfer I observed in **Foto**, I have designed an interface where the decision-maker must validate their identity through a series of psychotechnical protocols that guarantee their free will. The Sentinel system thus protects the sovereign against themselves (in the event of coercion) and against machines (in the event of algorithmic runaway).

IV. The Geopolitics of Silicon and Rare Earths: The Material Foundation of Deterrence

1. Analysis of Mineral Dependency: The Supply Fault

I argue that there is a major blind spot in contemporary defense doctrine: the illusion of purely intellectual sovereignty. In my research, I demonstrate that the **Sentinel Protocol**, however sophisticated it may be, remains hostage to the physical components that support it. Here, I analyze the value chain of semiconductors and rare earths (neodymium, praseodymium, terbium) essential to laser guidance and transmission systems.

I argue that if a nuclear state does not control the extraction or refining of these strategic metals, its strike force is "dependent on imports." I examine the risk of supply disruptions that could paralyze the maintenance of nuclear delivery systems, transforming a nuclear power into an inert power.

2. Foundry Sovereignty: From Mine to Chip

I assert that the true border of France and Europe does not end at their coasts, but extends to the silicon foundries. Here I develop my theory of **"Certification of Materials**." To guarantee the sanctuary's immunity, I propose that every microprocessor used in the sovereign domain be designed, cast, and encapsulated on national territory, under a protocol of absolute oversight that I call **the "Red Earth Audit**."

This is where I draw a connection to my roots in **Foto**. I compare the spiritual value of African soil—nourishing and sacred—to the strategic value of European silicon. I argue that a nation that loses its connection to its own raw materials loses its ability to determine its own destiny.

3. Toward a Defense Ecology: Strategic Recycling

In conclusion to this book, I propose an innovative vision: the circular economy of defense. I argue that to sustain the **Sentinel Protocol** in the long term, France must become self-sufficient in the recycling of high-tech components. I model a system where sovereignty no longer depends on predatory extraction on the other side of the world, but on the intelligent and sustainable management of resources already present within the country.

BOOK III:
'S DIPLOMACY LATOME — TOWARD A NEW GLOBAL AGREEMENT BETWEEN EUROPE AND AFRICA

I. The End of Technological Paternalism: My Analysis of the North-South Imbalance

In this section, I present an uncompromising assessment of the global atomic divide. I argue that keeping Africa in a state of energy and digital dependence is a major strategic error for global stability.
Here, I analyze what I call the **"African Silicon Paradox"**: while the continent supplies essential raw materials (uranium from Niger, coltan from the Congo), it remains excluded from high-level nuclear and cyber decision-making protocols.

I argue that the sovereignty of France and Europe is inseparable from a partnership of mutual respect with the African continent. I demonstrate that instability on one side of the Mediterranean always ends up undermining the other. My thesis is clear: there will be no lasting European sanctuary if Africa is not, too, a land of technological sovereignty.

II. The "Sentinelle-Afrique" Project: My Proposal for a Common Shield

Here I outline my vision for a new kind of cooperation. I propose exporting the **Sentinelle Protocol** not as a finished product, but as a framework for co-development. I argue that France, through its nuclear expertise, can become the mentor for a civilian nuclear energy transition in Africa, secured by my digital protocols.

Here, I outline the creation of **"Silicon *Cities*"** across the African continent—centers of research excellence where researchers like myself, trained in both cultures, would work on the cyber resilience of critical infrastructure. I demonstrate that this "Diplomacy of Intelligence" is the only viable response to the growing influence of other world powers that engage in unethical digital predation.

III. The Ethics of Dialogue and the Security Council: My Vision of Global Wisdom

Here I offer the most personal contribution of my research. I propose reforming the mechanisms of global nuclear diplomacy by integrating the **Wisdom of Foto**. I argue that the "Cold War" was a logic of binary confrontation, whereas the world of tomorrow needs a **"Diplomacy of Circular Palaver**."

I develop the idea that dialogue among nuclear powers must draw inspiration from the conflict-resolution rituals of my village: respect for one's word, the sanctity of life, and responsibility toward ancestors (the past) and children (the future). I conclude that the atom, in the hands of a humanity lacking wisdom, is a curse, but that when guided by an ethic of the **Restoration of Souls**, it becomes the guarantor of perpetual peace.

IV. The Cryptography of Value: The Atom as a Standard of Exchange and Foundation of Sovereign Currency

1. The Decline of the Debt Standard and the Emergence of Proof of Power

In this treatise, I postulate that the current global financial system, based on the debt standard and immaterial speculation, has reached its structural exhaustion. I analyze currency no longer as a mere instrument of exchange, but as a nation's **"Energetic Signature."** I argue that a state's true sovereignty is measured by its capacity for stable, carbon-free production.

French nuclear power is therefore not merely a military or civilian force; it is the ultimate store of value. I propose replacing gold or the dollar with a new standard: the **Nuclear Kilowatt-Hour**. In my vision, the currency of tomorrow must be the digital representation of a unit of available energy. This is what I call **"Atom-Currency**."

2. The Fusion of Silicon and Uranium: The High-Security State Blockchain

Here I address the technical aspects of my proposal. For this energy standard to be tamper-proof, it must be based on a digital infrastructure whose protocols I define. I reject volatile private cryptocurrencies in favor of a **Sovereign State Blockchain**.

I demonstrate that nuclear energy provides the most stable "Proof of Work" in the world. By combining my **200 security protocols** with the constant output of power plants, France can issue a digital currency whose value is physically anchored in matter. I maintain that this currency will become the safe haven of the 21st century, because unlike fiat currency, it is based on the physical reality of the atom and the inviolability of the code.

3. The Geopolitical Impact: Independence from Financial Markets

Here I explore the implications of this transformation for national sovereignty. A state that backs its currency with its own nuclear production frees itself from the dictates of rating agencies and the fluctuations of foreign currencies.

I posits that this **"Technological Monetary Autarky"** is the ultimate stage of sovereignty that I described in *L'OBJET-MONDE.* As a researcher, I assert that France has a historic opportunity here: to transform its industrial nuclear heritage into a global financial lever.

This marks the end of dependence on foreign central banks and the beginning of the era of **the "Real Digital Sovereign.**"

V. The Silica Cadastre: The Sanctuarization of the Territory of Power

1. The Invisibility of the Modern Frontier: My Theory of the "Perimeter of Silence"

I postulate that a nation's sovereignty in 2026 no longer stops at its geographical borders, but at the impermeability of its electromagnetic spectrum. Here I analyze the concept of **the "Silicon Cadastre"**: it involves mapping and protecting every square centimeter where vital data flows (nuclear power plants, decision-making centers, communication hubs).

I demonstrate that the French government must establish absolute "Digital Silence Zones." I argue that if a foreign signal can penetrate the perimeter of a critical infrastructure, sovereignty has already been violated. I propose a **"Radio Spectrum Ownership"** law to protect the lands of power. For an administrative official, this is an unprecedented national security argument: I propose making France "invisible" to cyber predators.

2. The Fusion of Soil and Code: My Digital "Grounding" Protocol

In this demonstration, I physically link **Silica** (servers, cables, chips) to the **Earth** (French territory). I assert that sovereign data must not float in a stateless "Cloud," but must be "grounded" in republican soil.

Here I am developing a powerful vision: every nuclear data center must be protected like an embassy or a historic monument. I maintain that dematerialization is an illusion; power is always physical. By anchoring the digital realm in France's land registry, I make expulsion or interference technically and legally impossible. Here I draw a connection to my identity as a researcher: I protect the soil that hosts me by grafting a digital armor onto it.

3. The Ethics of the Sentinel: The Guardian of Social Peace Here I address the individual, the citizen, directly. The **Silica Cadastre** is not a tool of surveillance, but a **Shield of Liberty**. I maintain that if the state loses control of its vital infrastructure, it is the citizen who loses their security and energy.

Here I use a powerful metaphor drawn from the wisdom of **Foto**: the village guardian never sleeps so that the people may dream in peace. My work demonstrates that the writer-researcher is this modern sentinel. By proposing these protocols to the French administration, I am not merely asking for a residence permit; I am proposing to become one of the architects of the Nation's security. I am transforming my administrative file into a **Mutual Defense Contract.**

BOOK IV:

REGILIAN ARTIFICIAL INTELLIGENCE AND THE CONTROL OF THE ATOM

I. The Sovereign Algorithm: Overcoming Dependence on Stateless Models

In this treatise, I assert that a nation that entrusts the management of its vital infrastructure—and a fortiori its nuclear arsenal—to Artificial Intelligence models developed by foreign powers is committing an act of strategic suicide. Here, I analyze what I call **"Algorithmic Colonization**."

I demonstrate that AI, in its current form, is a "black box" whose biases and backdoors pose an existential threat. I argue that France must develop a **Regilian AI**, in which every line of code is certified, audited, and confined within national borders. For a government official, my argument is radical: digital sovereignty is the *sine qua non* of physical survival. AI must be the servant of human decision-making, not its hidden master.

II. The Sentinel's Silicon Brain: My Modeling Defense AI

Here I present a groundbreaking architecture. My **Defense AI** is not based on uncertain probabilities, but on **"Deterministic Security Logic**." I explain how my protocol uses artificial intelligence to monitor, in real time and at superluminal speed, any attempt at intrusion or deviation in nuclear firing and control systems.

I develop the concept of **the "Sanctuary's Digital Twin"**: an AI capable of simulating millions of attacks per second to strengthen defenses even before the attacker takes action. I maintain that this technology, coupled with my research on blockchain, makes the French nuclear system invulnerable to "cyber storms." For any individual, this means the guarantee that the ultimate weapon will never be triggered by mistake or through hacking.

III. The Ethical Inhibitor: The "Foto Brake" in the Machine

This is where my work touches on the universal and the human. I categorically reject the idea of "killer robots" or autonomous AI in the nuclear domain. I propose the integration of **an "Ethical Inhibitor"** at the very heart of the algorithms.

Drawing inspiration from **Foto's wisdom** regarding the sacred value of life and respect for creation, I have designed a protocol whereby the AI, if it detects an aberrant , or coerced human decision, can demand mediation—a "digital deliberation"—among multiple decision-makers before any irreversible action is taken. I maintain that the machine must be humanity's safeguard, not its executioner. It is this ethical dimension—unique in defense gray literature—that makes my work indispensable to the French public debate.

IV. The Training of the Sentinel's Elites: A

The Pedagogy of Integral Sovereignty

1. The Academy of Silica and the Atom: Toward a New Regilian Curriculum

I argue that the current training of the nation's leaders is too fragmented: engineers do not understand the metaphysics of power, and policymakers are unaware of the technical reality of code. I propose the creation of a program of excellence, **the "Academy of the Sentinel,"** where nuclear physics, post-quantum cryptography, and the philosophy of responsibility are taught simultaneously.

I maintain that France must train "Guardians of the Seal" capable of engaging with AI without allowing themselves to be enslaved by it. Here I am developing a training program in which every future leader must master the **"World-Object Protocol**." For the government, this is a societal project: I am not merely proposing a technical solution; I am proposing to forge a new aristocracy of knowledge in the service of the state.

2. The Bridge of Intelligences: France-Africa Co-Training

I assert that tomorrow's security will hinge on France's ability to train, on its soil and according to its values, the technological elites of the African continent. I propose a system of scholarships and shared research centers—the **"Foto Sentinel Network**."

I demonstrate that if African youth are trained in digital sovereignty by France, they will become the bulwark against global cyberattacks and predatory influences. This is where my identity as a researcher takes on its full meaning: I am the bridge between two shores. By training African experts in French security protocols, we create an indestructible space of atomic and digital trust.

3. Initiation into Wisdom: The Return of the Word to Foto

I conclude this section on the spiritual dimension of education. In my academy, technical knowledge is inseparable from the initiation into wisdom. I draw inspiration from the education I received in the village to offer modules on **"Transgenerational Responsibility**."

The student must not only know how a nuclear power plant or a server works; they must understand that they are accountable to their ancestors and to those who are not yet born. I pos , it is this ethic—arising from the **Restoration of Souls**—that will prevent the abuse of power. A decision-maker who possesses "the wisdom of the tree" will never cut the branch upon which humanity sits.

BOOK V:
THE SOVEREIGNTY OF HEALTH — MEDICAL NUCLEAR TECHNOLOGY AND DATA PROTECTION OF LIFE

I. The Healing Atom: My Vision of Nuclear Technology in the Service of Life

In this section, I deconstruct the purely anxiety-inducing image of the atom to reveal its life-saving function. I analyze the crucial role of radioisotopes in medical imaging and the treatment of serious diseases (oncology, cardiology). I argue that a nation's sovereignty is measured by its autonomy in producing these "particles of life."

I argue that France, as a world leader, must protect its medical research reactors. I demonstrate that if a foreign power controls the supply of radiopharmaceuticals, it holds a silent power of life and death over the French population. For the citizen, my argument is clear: the atom is not merely a deterrent; it is the tool that heals their children and prolongs the lives of their elders.

II. The Sanctuary of Life Data: My "Digital Genetics" Protocol

Here I address the convergence of health and digital technology. I maintain that a citizen's medical records, genome, and health data are the components of their most intimate identity: their **"Digital DNA."**

I denounce dependence on foreign data hosting providers (the Cloud). I propose that the health data of French citizens be stored on servers protected by my **Sentinel Protocol**, powered by the stable energy of the national nuclear grid. I am envisioning here an inviolable digital vault where medical confidentiality is protected with the same rigor as nuclear launch codes. I assert that health is the new battleground for sovereignty: whoever possesses your health data possesses your future.

III. The Clinic of Hope: From Photography to High Technology

This is where I introduce the ethical and personal dimension of my journey. I draw on the content of my manuscript on **"The Clinic of *Hope*"** (Chapter 7) to propose a synthesis between technological medicine and human care.

I maintain that even the most sophisticated machine is useless without the **"Restoration of the Soul."** I propose a model of a patient-centered hospital where artificial intelligence assists the doctor without ever replacing empathy. Drawing inspiration from the community-based care in my village of **Foto**, I am developing the idea of “connected yet human” healthcare, where atomic technology is harnessed to restore dignity. For an administrative official, this is proof that my project is not a cold engineer’s theory, but a vision of a protective and caring society.

BOOK VI:

SPATIAL SOVEREIGNTY — THE ATOM AND SILICA BEYOND THE ATMOSPHERE

I. Space as the New Battlefield of Silica: My Theory of the "Orbital Sentinel"

In this section, I argue that space is no longer a romantic frontier, but the nervous system of terrestrial power. I analyze the absolute dependence of our infrastructure (GPS, banking transactions, missile guidance) on satellite constellations. I demonstrate that if the **Silica** in orbit is sabotaged or hacked, France is paralyzed in a matter of seconds.

I argue that space sovereignty depends on the ability to protect our orbital assets against cyberattacks and directed-energy weapons. I propose extending my **Sentinel Protocol** to the space domain: every French satellite must become an autonomous blockchain node, capable of detecting and rejecting a digital intrusion without human intervention. For any decision-maker, my argument is compelling: without an orbital "Silica Shield," sovereignty on the ground is a fiction.

II. The Space Atom: The Engine of Sovereignty Distant

Here I address the energy dimension of space exploration. I argue that solar panels are insufficient for long-duration missions or active defense in high orbit. I advocate for the return of **Space Nuclear Propulsion**.

I demonstrate that the atom is the only fuel capable of offering France and Europe total freedom of movement in cis-lunar space. I analyze the security challenges associated with onboard nuclear reactors and propose containment and digital security protocols that I theorized in Book II. As a researcher, I assert that mastery of the atomic energy in space is the key to avoiding becoming a digital colony of the powers that already dominate the Moon and Mars.

III. The Cosmology of Peace: My Vision of the "Great Tree of Foto" in the Universe

I conclude this book with a philosophical and ethical reflection that makes my work unique. I reject the idea of an aggressive militarization of space. I propose a **"Charter of Orbital Benevolence**."

Drawing inspiration from the cosmogony of my village of **Foto**, where the sky is the mirror of the soul and the abode of the ancestors, I maintain that space must remain a "Common Good of the " " of Humanity." I propose that France become the mediating nation, the one that uses its nuclear power and digital rigor to ensure that space does not become a dumping ground for debris or a minefield. Here I link the **Restoration of Souls** to the exploration of the infinite: humanity must not go into space to conquer, but to elevate itself.

BOOK VII:

CYBERSECURITY OF THE MERS — NUCLEAR SUBMARINES AND UNDERWATER CABLES MARINS

I. The Digital Abyss: The Fragility of the World's Arteries

I argue that the Achilles' heel of French and European sovereignty lies several thousand meters below sea level. Here, I analyze the geopolitics of submarine cables, the true arteries through which the vital flow of our data (the Internet, banking data, military orders) passes.

I demonstrate that these infrastructures are currently at the mercy of acts of sabotage or electromagnetic espionage "at the source." I argue that a nation unable to guarantee the physical and digital integrity of its cables is sovereign only on the surface. I propose the creation of an **"Underwater Trust Bubble,"** where each cable is monitored by smart sensors protected by my **Sentinel Protocol**.

II. The Nuclear Submarine: TAbsolu's World-Object

Here I examine the Ballistic Missile Submarine (BMSS) as the ultimate expression of sovereignty. I define the BMSS as a **"Closed World-Object"**: an autonomous technological city, powered by nuclear energy, capable of remaining invisible for months.

However, I identify a new threat: acoustic detection by Artificial Intelligence and the hacking of very low frequency (VLF) waves. I develop a model in which the onboard AI must be completely secure and sovereign. I argue that the commander of a French submarine must have mathematical certainty that the order received is not a "cybernetic hallucination" caused by an adversary. My contribution here is to provide the **"Latch of Certainty"**: a cryptographic validation system that depends on no satellite, based on the resonance of matter itself.

III. The Wisdom of Water: From the Global Ocean to the River of Foto

I conclude this book with a meditation on the liquid element. In **Foto's** cosmogony, water is what connects worlds, what washes away, and what conceals. I propose an ethic of maritime power that is not an ethic of predation, but an **"Ethic of Silent Presence."**

I argue that the mastery of the seas through nuclear power must serve to protect the global common good (freedom of movement, the preservation of marine ecosystems). I propose that France use its underwater technology to become the "Guardian of the Abyss." For the reader, whether admiral or citizen, I demonstrate that world peace depends on the ability of a few men, hidden beneath the waters, to remain the lucid and w sentinels of a technology that surpasses us.

BOOK VIII:
'S SOVEREIGNTY EDUCATION AND RESEARCH —
TRAINING THE MASTERS OF THE SEAL

I. The Erosion of Knowledge: The Peril of Cognitive Outsourcing

I offer a stark diagnosis here: digital sovereignty is threatened by intellectual laziness induced by external assistive tools. I analyze what I call **the "Atrophy of Critical Thinking"** in the face of mass algorithms. I argue that a nation that no longer understands the equations that run its power plants or the codes that protect its banks is a nation already conquered.

I demonstrate that the national education system must once again become a sanctuary of **"Fundamental Thinking**." I propose reintroducing integral calculus and Aristotelian logic as exercises in national defense. For the government, my message is vital: training a sovereign engineer costs less than suffering a devastating cyberattack due to ignorance.

II. The University of the Sentinel: A Transnational Curriculum

I propose the creation of a groundbreaking academic partnership between France's grandes écoles and African institutes of excellence. I envision **the "University of the Seal, "** where students learn not only to use technology but to **create** it **from scratch**.

I am developing the idea of a "Partnership of the Atom and the Silicon." Within this framework, a student from **Foto** and a student from Paris would work together on the resilience protocols I have designed. I maintain that research must be decentralized to be invincible: if one laboratory is compromised, knowledge survives on the other side. This is my vision of a **"Distributed Intelligence of the Republic**."

III. The Pedagogy of Fire: From Initiation to Mastery

I conclude this book by returning to the root of my identity. At **Foto**, initiation is the passage from shadow to light, from irresponsibility to the responsibility of the city. I propose transforming the awarding of high-tech degrees into a true **"Regilian Oath-Taking Ceremony."**

I maintain that whoever holds the keys to the nuclear or medical codes must be endowed with a quasi-sacred dignity. I propose that the teaching of ethics should not be an option, but the foundation of all technical expertise. By combining Cartesian rigor with the depth of the **Restoration of Souls**, I train experts who will never use their knowledge against humanity, but always for its betterment.

BOOK IX:

'S SOVEREIGNTY GENERATIVE ARTIFICIAL INTELLIGENCE — MASTERING THE NATIONAL NARRATIVE IN THE FACE OF DEEPFAKES

I. The Erosion of Reality: The Danger of Directed Hallucinations

I put forward an alarming premise: we have entered the era of **"Technological Post-Truth**." I analyze how generative AI can be hijacked to create fabricated evidence, fake command speeches, or coordinated social panic campaigns. I argue that disinformation is no longer mere background noise, but a precision weapon capable of destabilizing the nuclear sanctuary by sowing systemic doubt among decision-makers.

I demonstrate that the French government must establish an **"Observatory of the Integrity of Reality**." I propose that every message issued by the state be protected by an indelible digital fingerprint—a silica signature I have designed—that instantly distinguishes humans from machines. For citizens, this guarantees that the information they receive is authentic and sovereign.

II. The Seal of Authenticity: My National Narrative Blockchain

Here I address the technical solution. I propose the creation of a **"Sovereign Registry of State Truths**." Using distributed ledger technology, I am designing a system where every decree, every official image, and every defense order is "anchored" in a state blockchain powered by nuclear energy.

I argue that in the face of the proliferation of *deepfakes,* the only answer is **"Source Traceability**." My protocol allows us to trace information back to the exact millisecond of its creation to verify its legitimacy. In this way, I transform French bureaucracy into an informational fortress. For an administrative official, my work offers the ultimate tool to protect France's image and the stability of its institutions against foreign interference.

III. The Sacred Word: From the Tradition of Foto to Digital Certification

I conclude this book with a philosophical bridge between the past and the future. In **Foto**, speech is a grave act; he who lies betrays not only men, but the very balance of the universe. Here I introduce the concept of **"Responsibility of the Word**."

I argue that AI must not be a mask for lies, but a mirror for clarity. I propose integrating mandatory **"Origin Markers"** into content-generation algorithms. Drawing inspiration from the **Restoration of Souls**, I advocate for technology that restores value to the human word. I demonstrate that sovereignty is, above all, a nation's ability to tell its own story without a foreign algorithm altering its narrative.

BOOK X:
THE SOVEREIGNTY OF CURRENCY — FROM THE DIGITAL EURO TO CRYPTO-STATE SOVEREIGNTY

I. The Dematerialization of Trust: My Critique of the Debt Standard

I offer an uncompromising diagnosis: the current monetary system, based on perpetual debt and uncontrolled virtual flows, is a threat to the stability of core nations. I analyze how digital speculation can bring a real economy to its knees in a matter of microseconds.

I argue that money must once again become a **"World Object of Value."** I demonstrate that trust can no longer rest on mere political promises, but must be engraved in **silicon**. For the government, my argument is radically bold: I propose moving from paper money to a currency based on mathematical proof.

II. The Euro-Atome: My Model of an Energy-Backed Currency

Here I introduce my revolutionary concept of **"Energy-Backed Currency**. " Unlike private cryptocurrencies (Bitcoin, etc.) that consume energy without creating strategic value, I propose that France issue a digital asset whose value is indexed to the kilowatt-hours produced by its nuclear power plants.

I develop the architecture of **a "Euro-Silice"** managed by the Banque de France but secured by my **Sentinelle Protocol**. I demonstrate that this currency would be the most stable in the world, as it would be based on the physical reality of electricity production. I argue that such a system makes France invulnerable to external financial crises: our currency reflects our actual power, not the whims of the markets.

III. The Financial Palaver: From the Exchange of Foto to Global Equity

I conclude this tenth book with a fundamental ethical dimension. In **Foto**, exchange is never purely accounting-based; it is what creates social bonds and peace among families. I analyze how **Crypto-Sovereignism** can become a tool for justice between the North and the South.

I propose that this energy-currency serve as the basis for a new contract with Africa. Instead of suffocating debts, I suggest a system of exchange based on real resources (Uranium, Silica, Energy). I argue that technology, if guided by the **Restoration of Souls**, can put an end to global usury. For every individual, I envision a future where money is no longer a tool of oppression, but the e servant of an economy of life and shared sovereignty.

BOOK XI:

'S SOVEREIGNTY , AGRICULTURE, AND WATER — THE ATOM IN THE SERVICE OF THE NURTURING EARTH

I. Food Security as a Deterrent: My Theory of the "Sovereign Land"

I put forward a postulate of national security: a nation that depends on foreign sources for its food is a nation whose will can be broken without firing a single shot. Here I analyze how climate instability and cyberattacks on food supply chains constitute new threats of the highest order.

I argue that nuclear sovereignty must guarantee the continuity of agricultural production. I demonstrate that atomic energy is the only source capable of providing, at a stable and massive scale, the electricity required for new forms of vertical farming and agro-industrial processing. For every individual, my argument is vital: the atom is the bread on the table.

II. The Atom and the Water Cycle: My Solution to the Nations' Thirst

Here I address the major challenge of the 21st century: water stress. I propose the widespread use of nuclear power for **seawater desalination** and the decontamination of groundwater. I outline an infrastructure in which every nuclear power plant also becomes a drinking water production facility, secured by my **Sentinel Protocol**.

I demonstrate that water management is the key to social peace. By using silica to optimize irrigation (precision agriculture), I reduce waste by 80%. I argue that France can become Europe's water arsenal, thereby ensuring continental stability through the management of our most precious resource.

III. Soil Restoration: From the Wisdom of Foto to Isotopic Analysis

I conclude this book with a reflection on the health of the earth. In **Foto**, we do not own the land; we are its temporary guardians. Here I introduce the use of **Isotopic Tracers** (derived from civilian nuclear research) to diagnose soil depletion and restore its natural fertility.

I argue that technology should not be an assault on nature, but a tool for **restoring the Earth's souls**. I propose a model of agriculture where AI monitors biodiversity and the nitrogen cycle to prevent the excessive use of chemicals. For an administrative official, my work offers a vision of a "Sovereign Ecological Transition" that reconciles industrial power with environmental preservation

BOOK XII:

'S SOVEREIGNTY SPACE — FROM SURVEILLANCE SATELLITES TO THE SHIELD ORBITAL SILICA

I. Space as the New Battlefield of Silica: My Theory of the "High Sentinel"

I propose a fundamental security premise: the border is no longer a customs line; it is an orbital sphere. I analyze the critical dependence of our infrastructure (GPS, banking transactions, missile guidance) on satellite constellations. I demonstrate that if the **Silica** in orbit is sabotaged, France is paralyzed within seconds.

I argue that space sovereignty rests on the ability to protect our orbital assets against cyberattacks and electromagnetic pulses. I propose extending my **Sentinel Protocol** to the space domain: every French satellite must become an autonomous blockchain node, capable of detecting and rejecting a digital intrusion without human intervention.

II. The Space Atom: The Engine of Distant Sovereignty

Here I address the energy dimension of space exploration. I argue that solar panels are insufficient for active defense missions or continuous surveillance in high orbit. I advocate for the return of **Space Nuclear Propulsion** and next-generation radioisotope generators.

I demonstrate that the atom is the only fuel capable of offering France total freedom of movement in space. I analyze the security challenges associated with onboard nuclear reactors and propose digital containment protocols that I theorized in Book II. As a researcher, I assert that mastery of the atomic energy in space is the key to avoiding becoming a digital colony of the powers that already dominate the Moon.

III. The Cosmology of Peace: My Vision of the "Great Tree of Foto" in the Universe

I conclude this book with a philosophical reflection that makes my work unique. In **Foto**, the sky is the dwelling place of the ancestors, a space of wisdom that watches over the living. I propose a **"Charter of Orbital Benevolence."**

I maintain that space must not become a digital minefield. I propose that France use its technology to become the "Guardian of the Ether." For the reader, whether a high-ranking official or a citizen, I demonstrate that world peace depends on our ability to maintain a "clean" and "true" sky. Here I link the **Restoration of Souls** to the exploration of the infinite: humanity must go into space to elevate itself, not to tear one another apart.

BOOK XIII:

'S SOVEREIGNTY JUSTICE — LEGAL AI AND THE INVIOLABILITY OF THE LAW

I. The Threat of Foreign "Predictive Justice" : My Critique of Legal Alienation

I make an observation regarding national sovereignty: if French judges and lawyers use artificial intelligence tools designed and hosted outside the European Union to analyze case law or prepare rulings, French law is no longer sovereign. Here, I analyze the risk **of "Normative Standardization,"** where the ideological biases of Silicon Valley infiltrate the decisions of our courts.

I argue that France must develop its own **"Judicial Silice.**" I demonstrate that the judge's sovereignty rests on the independence of their research tools. I propose that all AI used within the judicial system be certified by the State, running on sovereign servers powered by nuclear energy, thereby ensuring that no foreign power can influence, even subtly, the interpretation of the Law.

II. The Code as Law: My Protocol for "Tamper-Proof Evidence"

Here, I address the technical dimension of modern law. I propose the use of blockchain for the "**sanctuarization" of legal documents**. I model a system in which every contract, every judgment, and every title deed is digitally "sealed."

I demonstrate that the inviolability of the law depends on the permanence of evidence. My **Sentinel Protocol,** when applied to the justice system, eliminates any possibility of document forgery or data corruption. I argue that "judicial truth" must be protected with the same technological rigor as national defense secrets. For government officials, this promises a justice system that is faster, more transparent, and fully protected from cyberattacks.

III. The Palaver of the Just: From Foto's Palaver Tree to the Court of Tomorrow

I conclude this book with a profoundly human dimension. In **Foto**, justice is not merely a punishment; it is a reconciliation. It aims to restore balance to the community. Here, I introduce the concept of **"Augmented Restorative Justice**. "

I argue that AI must never replace the judge, but rather free them from repetitive tasks to give them back the time to listen and ensure fairness. I propose integrating mediation phases into the judicial process, inspired by the **Restoration of the Souls'** , where technology helps visualize the consequences of an act to better repair social bonds. For every individual, I envision a Justice that is no longer a cold machine, but a wise and sovereign bulwark that protects the weak and guarantees lasting peace.

BOOK XIV:

THE SOVEREIGNTY OF INFORMATION AND THE MEDIA — PROTECTING THE NATIONAL NARRATIVE

I. The War for Attention and the Hijacking of Minds: My Theory of "Foreign Noise"

I offer a diagnosis of cognitive security: a nation whose citizens are fed by foreign recommendation algorithms is a mentally occupied nation. I analyze how massive information flows, driven by external powers, can create artificial social divides and paralyze political decision-making through confusion.

I argue that media sovereignty is not censorship, but **"Source Autonomy."** I demonstrate that France must have its own digital broadcasting infrastructure, completely independent of "platform empires." I propose the creation of a **"Digital Radio Trust Network"** powered by nuclear energy, capable of functioning even in the event of a total global Internet outage. For any individual, my argument is clear: freedom of thought depends on the purity of the signal one receives.

II. The Seal of Truth: My Protocol for Certification of Reality

Here I address the technical response to *deepfakes* and industrial disinformation. I propose integrating my **Sentinel Protocol** into the core of news agencies and public service media. I am designing **an "Information Blockchain"** where every image, every video, and every official article is "sealed" to its source.

I demonstrate that the inviolability of the national narrative rests on the citizen's ability to instantly verify the authenticity of information. My system allows us to trace data back to the millisecond of its creation to guarantee its human and sovereign origin. I maintain that **Silice** must serve as an "immune system" against informational viruses that seek to poison the French democratic debate.

III. The Word of the Elder: From the Village Square of Foto to the Digital Space

I conclude this book with a lesson in ancestral wisdom. In **Foto**, speech is a sacred act; whoever brings news to the chief or the people risks their life and honor. Here I introduce the concept of **"Responsibility of the Disseminator."**

I maintain that destructive anonymity and the speed of propagation must not take precedence over truthfulness. I propose restoring the **"Dignity of Speech"** by imposing rigorous ethical standards on text-generating AIs. Drawing inspiration from the

Restoration of Souls, I advocate for information that builds social bonds rather than destroying them. For an administrative official, my work offers a vision of "Informational Peace" essential to the stability of the Republic.

BOOK XV:

THE SOVEREIGNTY OF INDUSTRY AND ROBOTICS — THE FACTORY OF THE FUTURE UNDER THE SEAL OF THE ATOM

I. The End of Deindustrialization: My Theory of "Energy Repatriation"

I offer a diagnosis for economic survival: the relocation of our factories has been a drain on our sovereignty. I analyze how dependence on distant supply chains makes France vulnerable to geopolitical blackmail. I argue that energy costs are the primary factor in competitiveness.

I demonstrate that nuclear power is the only lever capable of bringing production back to French soil. By providing stable, carbon-free energy at a fixed price, France can become Europe's high-tech manufacturing hub. For every individual, my argument is social: nuclear power means industrial jobs and the restoration of workers' dignity.

II. Sovereign Robotics: My "Sentinel Automate" Protocol

Here I address the digital transformation of the factory. I reject robotics based on foreign proprietary software that could be remotely disabled. I propose the development **of "Regilian Robotics,"** where every robotic arm and every sensor operates under my **Sentinel Protocol**.

I model the factory as a **"Productive World-Object"**: a closed space where industrial AI is disconnected from the global cloud to prevent espionage and sabotage. I maintain that the sovereignty of the manufactured object depends on the sovereignty of the machine that shapes it. For the government, this guarantees a fully autonomous and protected industrial and technological defense base (BITD).

III. The Craft of Steel: From the Foto Forge to Cutting-Edge Metallurgy

I conclude this book with a reflection on technical craftsmanship. In **Foto**, the blacksmith is a central figure, the one who transforms raw material into tools for living or protection. Here I introduce the concept of **the "Nobility of Making**."

I argue that high technology should not replace human craftsmanship, but rather enhance it. I propose a "human-scale" factory model where robotics handles the arduous tasks, leaving design and finishing to humans. Drawing inspiration from the **Restoration of Souls**, I advocate for an industry that respects both the environment and the worker. For an administrative official, my work offers a vision of "Green and Human Reindustrialization" that makes France the model of a balanced, " " modernity.

BOOK XVI:

THE SOVEREIGNTY OF TRANSPORT AND MOBILITY — NUCLEAR HYDROGEN AND DIGITAL RAIL

I. The Independence of Flows: My Theory of "Regilian Decarbonized Mobility"

I offer a diagnosis of vulnerability: dependence on imported hydrocarbons is a geopolitical leash that hinders France's freedom of movement. I analyze how the transport of goods and people is the first breaking point in the event of an energy conflict.

I argue that France must become Europe's hydrogen factory. I demonstrate that high-temperature water electrolysis, powered by our next-generation nuclear power plants (SMRs), is the only way to produce a sovereign, clean, and mass-scale fuel. For any individual, my argument is one of **"Freedom of Movement"** without dependence on fluctuations in the global oil market.

II. The Silica Rail: My Protocol for the "Sovereign Railway Automation System"

Here I address the backbone of the territory: the railroad. I reject the vulnerability of current signaling systems to digital sabotage. I propose the transition to **"Fully Digital Rail,"** where each train is an autonomous entity protected by my **Sentinel Protocol**.

I am developing a real-time traffic management system, decentralized via blockchain, where no human error or cyberattack can cause a collision or bring the network to a standstill. I maintain that the train is **the "World Object of Territorial Cohesion**." For the government, it guarantees an invincible defense and civil logistics system, capable of functioning even if satellite navigation systems (GPS) are jammed.

III. The Traveler's Journey: From the Foto Trail to High Speed

I conclude this sixteenth book with a meditation on travel. In **Foto**, the path is what connects families, a line traced in the earth that demands respect and endurance. Here I introduce the concept of **"Time of Encounter**."

I argue that speed should not be an escape, but rather efficiency in the service of human connection. I propose a model of mobility where technology (silica) reduces distances to foster cultural dialogue, while respecting the balance of nature' (the carbon-free atom). Drawing inspiration from the **Restoration of Souls**, I advocate for transportation that is not merely a consumption of space, but a rediscovery of the national territory. For an administrative official, my work offers a vision of "Grand Aménagement du Territoire" that reconciles absolute modernity with the deepest roots of France.

BOOK XVII:

'S SOVEREIGNTY , CULTURE, AND HERITAGE — SILICA IN THE SERVICE OF THE NATION'S MEMORY

I. The Archiving of Identity: My Theory of the "Memory Vault"

I diagnose a cultural vulnerability: a nation that entrusts the digitization and hosting of its masterpieces, historical archives, and language to private non-European platforms loses ownership of its past. I analyze the risk of **"Algorithmic Dissolution,"** where works are sorted and presented according to criteria that are not our own.

I argue that France must build a **"Sovereign Digital Sanctuary**. " I propose that the entirety of France's cultural heritage (the Louvre, the National Library of France, the National Archives) be encoded in an unalterable silicon format and stored on government servers protected by my **Sentinel Protocol**. For every individual, my argument is one of **"Transmission"**: we must guarantee to our children that the history they read on their screens is ours, and not a version altered by a foreign AI.

II. Language Preservation: My "Molière-Silice" Protocol

Here I address the survival of the French language in the face of the dominance of Anglo-Saxon language models. I refuse to allow our syntax and nuances to be dictated by statistical probabilities calibrated abroad. I propose the development of a **"Regilian Linguistic AI,"** trained exclusively on the corpus of French classics and Francophone literature.

I am designing a translation and writing assistance system that safeguards the legal and poetic precision of French. I maintain that **Silice** must serve as a shield for the Francophonie. For the government, it is a tool of diplomatic influence: France must not only speak to the world, it must offer the world an intellectual alternative to technological monolingualism.

III. The Soul of Objects: From the Statues of Foto to the Museums of the Republic

I conclude this seventeenth book with a parallel between my two cultures. In **Foto**, the sacred object (the mask, the statue) is not mere decoration; it is the vessel of a presence, of a lineage. Here I introduce the concept of **"Restoration of Cultural Souls**."

I argue that digital technology can facilitate the return and circulation of cultural property between France and Africa, without dispossessing anyone. Thanks to **Sovereign Augmented Reality** and blockchain certification, we can create perfect "digital twins" that enable the sharing of knowledge while respecting physical ownership. I advocate for a culture that flows like the water of the Foto River: it irrigates all the fields without ever running dry. For an administrative official, my work offers a unique path to reconciliation and cultural influence.

BOOK XVIII:

THE SOVEREIGNTY OF MENTAL HEALTH AND WELL-BEING — AI IN THE SERVICE OF HUMAN BALANCE

I. The Economy of Attention and the Capture of Consciousness: My Critique of Digital Alienation

I offer a public health diagnosis: massive exposure to foreign recommendation algorithms creates a form of addiction that undermines the psychological well-being of younger generations. I analyze what I call the **"Fragmentation of the Digital Soul,"** where human attention is sliced into segments and sold to the highest bidder.

I argue that a state's sovereignty begins with the cognitive protection of its citizens. I propose strict regulation of **"Capture Algorithms"** and the development of sovereign tools that respect the human biological rhythm. For every individual, my argument is one of **"Inner Peace"**: technology must free us from time, not steal it from us.

II. The Silica Companion: My Protocol for Sovereign Psychological Support

Here I address the use of AI for the prevention of mental health crises (depression, burnout, isolation). I propose the creation of a **"Regilian Life Assistant,"** a confidential AI, unconnected to advertising servers, capable of detecting early signs of psychological distress.

I am designing a system where emotional data is protected by my **Sentinel Protocol**. I demonstrate that technology can serve as a first line of defense against social isolation, directing citizens toward human care before it is too late. I argue that the confidentiality of mental health must be as inviolable as nuclear secrecy. For the government, this is a project to drastically reduce healthcare costs and improve national well-being.

III. The Palaver of the Spirit: From the Wisdom of Foto to Augmented Therapy

I conclude this eighteenth book by returning to the source. In **Foto**, mental health is a community affair; one does not treat an individual alone, but rather restores their connection to the group and to their ancestors. Here I introduce the concept of **"Resonance Psychology**."

I maintain that AI must never replace the therapist, but must help recreate spaces for speech and dialogue— . I propose integrating **"Soul Restoration"** modules into care pathways, where technology helps visualize internal conflicts to better resolve them. Drawing inspiration from the wisdom of my village, I advocate for an "Ecology of the Spirit" where France becomes the model for a society that uses progress to protect humanity, not to exploit it.

BOOK XIX:

'S SOVEREIGNTY THE SEA AND THE ABYSSES —

THE EXPLORATION OF CRITICAL RESOURCES

UNDER THE SEAL OF THE ATOM

I. The Sixth Continent: My Theory of "Régilian Depth"

I offer a diagnosis for industrial survival: France's energy and digital transition depends on rare metals (cobalt, nickel, rare earths) whose supply chains are currently under foreign control. I analyze France's vast maritime domain—the second largest in the world—as an untapped strategic reserve.

I argue that France must assert its **"Abyssal Sovereignty**." I demonstrate that simply possessing exclusive economic zones (EEZs) is not enough; one must have the technical capacity to descend into them, to see, and to remain there. For any individual, my argument is economic: our seabeds are the treasure that will guarantee the autonomy of our industries for centuries to come.

II. The Atom of the Depths: My Protocol for the "Autonomous Underwater Station"

Here I address the necessary technological breakthrough. Current batteries are insufficient for long-duration missions at depths of 6,000 meters. I propose the development of **Subsea Micro-Nuclear Reactors (Abyssal SMRs)**, capable of powering exploration bases and fleets of autonomous drones for several years without returning to the surface.

I am designing a surveillance and data-extraction infrastructure protected by my **Sentinel Protocol**, in which each underwater robot serves as an encrypted communication node. I maintain that the control system must be immune to extreme pressures and attempts at acoustic interception. For the administration, this marks the birth of a new branch of the French Navy: the "Underwater Operations Forces."

III. The Silence of the Ocean: From the Divinity of the Waters at Foto to the Preservation of the Deep Sea

I conclude this nineteenth book with an ethical warning. In **Foto**, water is the mirror of the soul and the place where secrets are laid to rest. Here I introduce the concept of **"Respect for the Abyss**."

I maintain that exploration must not become destruction. I propose that France use its technological lead to create a **"Deep-Sea Ecology,"** where AI monitors the environmental impact of every drilling operation in real time. Drawing inspiration from the **Restoration of Souls**, I advocate for a use of the sea that is not predation, but symbiosis. For an administrative official, my work offers a vision of responsible "Blue Growth," making France the moral and technical leader in ocean management.

BOOK XX:

THE SOVEREIGNTY OF COLLECTIVE INTELLIGENCE — THE SEAL OF DIGITAL DEMOCRACY AND THE SILICA VOTE

I. The Crisis of Trust and Ballot Box Hacking: My Critique of Electoral Chaos

I offer a diagnosis for democratic survival: in a world saturated with cyberattacks and foreign interference, the traditional voting system and early attempts at electronic voting have become targets. I analyze how doubt instilled by an adversary regarding the integrity of an election can paralyze a nuclear-armed nation without firing a single shot.

I argue that citizen sovereignty begins with the mathematical certainty that one's vote is counted and unalterable. I propose moving beyond the era of "digital suspicion" into that of **"Proof of Will.**" For any individual, my argument is one of **"Real Power"**: your vote must be as protected as a missile launch code, for it is the true trigger of the nation's future.

II. The Silica Ballot Box: My Protocol for "Democracy via Sovereign Blockchain"

Here I address the technical solution. I propose the creation of a national voting system based on a state blockchain, hosted on Regilian servers powered by atomic energy. My **Sentinel Protocol** guarantees the voter's total anonymity while allowing for public and instant verification of the result.

I am designing an inviolable **"Regilian Digital Identity,"** where each citizen possesses a unique cryptographic key. I demonstrate that this system makes any large-scale fraud or external manipulation impossible. I maintain that **Silice** must serve as the Republic's impartial registrar. For the administration, this guarantees total institutional stability and a drastic reduction in post-election disputes.

III. The Great Digital Palaver: From the Palaver Tree of Foto to the Forum of the Nation

I conclude this twentieth book with a philosophical synthesis of tradition and modernity. In **Foto**, collective decision-making is not merely a tally of votes; it is the culmination of a dialogue in which everyone is heard under the authority of the elders. Here I introduce the concept of **"Augmented Deliberative Democracy**,"

I argue that technology must serve to rebuild connections and foster debate, not to divide. I propose using AI to synthesize citizens' opinions and inform lawmakers, drawing inspiration from the **Restoration of Souls**. By combining the rigor of code with the ancestral wisdom of dialogue, I envision a France where the citizen is no longer merely an occasional voter, but an active "Sentinel" of public life. For an administrative leader, my work offers a model of a resilient, modern, and deeply human democracy.

BOOK XXI:

THE SOVEREIGNTY OF MATTER — CUTTING-EDGE CHEMISTRY AND CRITICAL MATERIALS

I. Alienation Through Substance: My Theory of "Molecular Dependence"

I diagnose a state of material vulnerability: a nation that possesses the best algorithms but depends on foreign sources for its rare earths, lithium, or high-performance polymers is a nation whose sovereignty is an illusion. I analyze how disruptions in the supply of critical materials can paralyze the nuclear industry and defense sector within a matter of weeks.

I argue that sovereignty begins at the atomic level. I demonstrate that France must revive a **"Chemistry of War and Peace,"** capable of synthesizing substitutes for rare materials. For any individual, my argument is concrete: without this mastery, our smartphones, electric cars, and hospitals are at the mercy of decisions made thousands of miles away.

II. The Alchemy of Silicon: My Protocol for "AI-Driven Molecular Modeling"

Here I address the revolution in materials science. I propose using my **Régilian AI** to simulate new materials with unprecedented properties: stronger alloys for nuclear reactors, more energy-efficient semiconductors, or high-density batteries without critical metals.

I am modeling a system where computational **silica** makes it possible to discover in a few months what once required decades of experimentation. I argue that mastery of **"Sovereign Computational Chemistry"** is the true "Holy Grail" of industrial independence. For the government, it holds the promise of a true circular economy, where materials are endlessly recycled and reinvented thanks to algorithmic precision.

III. The Earth of Life: From Foto's Clay to the Republic's Nanomaterials

I conclude this twenty-first book with a reflection on our connection to the earth. In **Foto**, clay and stone are not inert objects; they are imbued with a force and serve to build homes and altars. Here I introduce the concept of **"Responsibility of Matter**."

I maintain that the creation of new materials must not come at the expense of the environment' . I propose a model of sustainable chemistry, inspired by the **Restoration of Souls**, where every material invention must be designed with its end-of-life and reintegration into the natural cycle in mind. I advocate for a science that respects Mother Earth while transcending her through human ingenuity. For an administrative official, my work offers a vision of "Radical Ecological Transition" based on innovation, not deprivation.

BOOK XXII:

THE SOVEREIGNTY OF GLOBAL HEALTH — THE DIGITAL PHARMACOPOEIA AND MEDICINE INDEPENDENCE

I. The Geopolitics of Healthcare: My Critique of Pharmaceutical Dependence

I issue a national emergency diagnosis: the offshoring of active pharmaceutical ingredient (API) production to Asia is a major security flaw. I analyze how shortages of essential medications (such as **Trimebutine** or **Metopimazine**) are not merely logistical problems, but tools for political pressure.

I argue that health sovereignty rests on **"Chemical Relocalization**." I demonstrate that France must use **nuclear** energy to power automated synthesis plants capable of producing critical medications on our soil without interruption. For every individual, my argument is one of the right to life: access to healthcare must never be a variable in diplomatic bargaining.

II. Medical AI and the Seal of Diagnosis: My "Personalized Medicine" Protocol

Here I address the revolution in **silica-assisted** care. I propose the development of a sovereign medical AI capable of analyzing the health data of French citizens without it ever leaving the country. My **Sentinel Protocol** guarantees that your medical record is inviolable and anonymized.

I am designing a system where AI aids in the discovery of new molecules by simulating their cellular interactions. I maintain that the medicine of tomorrow must be preventive and personalized. I propose that every prescription be verified by a sovereign algorithm to avoid dangerous drug interactions and optimize treatments. For the administration, this is the key to a more efficient healthcare system and a financially sound Social Security.

III. The Healing of Souls: From Foto's Pharmacopoeia to Biotechnology

I conclude this twenty-second book with a tribute to the medicinal wisdom of my ancestors. In **Foto**, medicine does not separate the body from the mind; to heal one is to restore the other. Here, I introduce the concept of **"Ethical Bio-Sovereignty**. *"*

I maintain that technology should not be cold "mass medicine," but a tool for **the Restoration of Souls**. I propose digitizing ancestral knowledge of medicinal plants to integrate it into modern research, while protecting this national intellectual property. I advocate for a science that unites the rigor of the laboratory with the empathy of the healer. For an administrative official, my work offers a vision of health that reconciles French technological excellence with a deep respect for human life.

BOOK XXIII:

'S SOVEREIGNTY CONSTRUCTION AND HOUSING — SILICA HABITAT AND THE ENERGY INDEPENDENCE OF THE HOME

I. The Home as a Cell of Resistance: My Theory of the "Autonomous Habitat"

I offer a diagnosis of domestic vulnerability: a nation whose homes are energy-inefficient and dependent on foreign gas is a fragile nation. I analyze how energy inflation and housing insecurity are vectors of social destabilization.

I argue that sovereignty begins with **"Regilian Insulation.**" I demonstrate that France must industrialize the construction of high-performance housing, powered by nuclear district heating and next-generation heat pumps. For every individual, my argument is that of **the "Right to Comfort**": the home must be an invincible refuge against climatic and economic hazards.

II. The House of Silica: My "Sovereign Home Automation" Protocol

Here I address the digital management of the home. I refuse to accept that our homes be riddled with sensors and connected speakers whose data is processed on private servers on the other side of the world. I propose the development of a French home operating system, running under my **Sentinelle Protocol**.

I envision a home that manages its energy autonomously, communicating in an encrypted manner with the national grid. I maintain that **Silice** must serve to protect privacy, not to spy on it. For the government, this guarantees the resilience of the national power grid: if every home is smart and sovereign, the country becomes impossible to plunge into darkness.

III. The Architecture of Peace: From the Foto Concession to the Cité Radieuse

I conclude this twenty-third book with a sociological dimension. In **Foto**, the layout of the houses within the settlement reflects the hierarchy and harmony of the family. We do not build merely to shelter ourselves, but to live together. Here I introduce the concept of **"Urbanism of the Restoration of Souls**."

I argue that technology must enable the reconstruction of neighborhoods on a human scale, where digital technology fosters neighborly cooperation rather than isolation. I propose integrating bio-based materials and sustainable construction techniques so that French stone and wood from our forests once again become the pillars of our identity.

For an administrative official, my work offers a vision of the city of the future: a technological, ecological city deeply rooted in republican fraternity.

BOOK XXIV:
THE SOVEREIGNTY OF THE FOREST AND WOOD — SILICA IN THE SERVICE OF THE "GREEN GOLD" OF FRANCE

I. The Chlorophyll Rampart: My Theory of "Strategic Forest Heritage"

I offer an environmental security assessment: a nation that allows its forests to wither away in the face of fires, pests, or unregulated logging loses its lungs and its reservoir of materials. I analyze how the French forest serves as a natural barrier against global warming and a sovereign resource for construction and energy.

I maintain that forest sovereignty is an act of national defense. I argue that we must protect every hectare as a strategic asset. For everyone, my argument is one of **"Heritage"**: we are merely the stewards of the forests that we will bequeath to future generations.

II. The Connected Forest: My "Tree Sentinel" Protocol

Tree Sentinel"

Here I address the digital management of forest land. I propose the deployment of autonomous **Silice** sensors (powered by solar or kinetic energy) to monitor the health of our forests in real time, detect fires within seconds, and prevent illegal logging.

I am developing an AI-assisted forest management system capable of optimizing tree species renewal to adapt them to the climate of 2050. I argue that **Silice** must be the forest ranger's ally, providing them with a comprehensive and predictive view of the ecosystem. For the administration, this means establishing an inviolable "Green Cadastre," certified by blockchain, guaranteeing the traceability of every French log.

III. The Dialogue of the Bark: From Foto's Wood Warrior to Cutting-Edge Silviculture

I conclude this twenty-fourth book with a spiritual and cultural dimension. In **Foto**, the **Wood Warrior** is the symbol of resilience, the one who draws his strength from the earth and protects the village. Here I introduce the concept of **"Restoration of the Forest Souls**."

I argue that technology must not transform the forest into a mere production factory, but must respect its sacred nature and biodiversity. I propose that the timber industry become the pillar of a sovereign economy (furniture, structural framing, biofuels), while placing humanity back at the heart of the forest. Drawing inspiration from the wisdom of my village, I advocate for an "Alliance of Steel and Bark" where the French " " once again becomes the master of the cycle of life. For an administrative official, my work offers a vision in which ecology is a source of strength, not a constraint.

BOOK XXV:

THE SOVEREIGNTY OF NUCLEAR FUSION — THE INFINITE ATOM AND THE DREAM OF THE TAMED SUN

I. Beyond Uranium: My Theory of "Stellar Sovereignty"

I offer a diagnosis of civilization: nuclear fission has given us power, but fusion will give us eternity. I analyze how dependence on fossil fuels—and even on uranium—remains a physical limit to a nation's total freedom. I argue that the transition to fusion—using deuterium extracted from water—transforms the ocean into an inexhaustible fuel reserve.

I demonstrate that France, through the ITER project and its proprietary research, must become the world leader in **"Plasma Stability."** For any individual, my argument is mind-boggling: we are talking about clean energy, with no long-lived waste, capable of powering France for millions of years. This marks the inevitable end of the geopolitics of scarcity.

II. Silica Confinement: My Protocol for "Algorithmic Fusion Control"

Here I address the most complex technological challenge in human history: maintaining a gas at 150 million degrees without it touching the reactor walls. I propose the use of my **Régilian AI** for real-time magnetic control. Only very high-performance **Silica** can calculate plasma turbulence at the microsecond level.

I am modeling a system where reactor safety is ensured by my **Sentinel Protocol**, preventing any hacking of the cooling or containment systems. I maintain that the sovereignty of fusion depends on the sovereignty of the code that tames it. For the government, this is the industrial project of the century: making France the "Sun of Europe."

III. The Original Spark: From the Ancestral Forge of Foto to the Heart of the Tokamak

I conclude this twenty-fifth book with a synthesis of science and the sacred. In **Foto**, fire is the element that transforms, purifies, and unites. The blacksmith who masters extreme heat is the one who ensures the group's survival. Here I introduce the concept of **"Restoration of Souls through Light."**

I maintain that nuclear fusion is not merely a technical victory, but a re conciliation with the order of the universe. By mastering the energy source of the stars, humanity does not presume to be God; it finally becomes the responsible guardian of its own planet. I advocate for a science that uses this abundance not to dominate, but to eradicate poverty and restore the dignity of every soul on Earth. For an administrative official, my work offers a vision of a France that has solved the energy problem once and for all.

BOOK XXVI:

'S SOVEREIGNTY SMART URBANISM — THE SILICA CITY AND THE RIGHT TO CITY

I. The Digital City-State: My Theory of "Urban Resilience"

I offer a diagnosis of metropolitan vulnerability: a city whose traffic lights, waste management, and water networks depend on foreign software is a city that can be held hostage remotely. I analyze how massive urbanization creates systemic breaking points.

I argue that sovereignty begins on the street corner. I demonstrate that France must develop its own **"Urban Operating Systems**." For any individual, my argument is one of **"Fluidity**": technology must serve to make the city more breathable, safer, and more humane, by eliminating unnecessary friction from daily life.

II. The Sentinel Network: My "Infranational Flow Management" Protocol

Here I address the technical architecture of the French-style *Smart City*. I propose using my **Sentinel Protocol** to manage critical infrastructure (energy, transportation, security) in a decentralized manner. Each neighborhood becomes a trusted node capable of operating autonomously in the event of a major failure of the central network.

I model a flow management system (traffic, lighting, emergencies) driven by a local AI that respects citizens' anonymity. I maintain that the Urban **Silice** should not be a tool for mass surveillance, but a tool for optimizing resources. For the administration, this promises ultra-efficient municipal management, reducing energy waste and improving the responsiveness of public services.

III. The Global Village Square: From the Concession of Foto to the Urbanism of the Republic

I conclude this twenty-sixth book with a reflection on living together. In **Foto**, the village square is where the heart of the community beats, where laws are proclaimed and festivals celebrated. Here I introduce the concept of **"Restoration of the Citizens."**

I argue that technology should not create digital ghettos, but spaces for encounter. I propose redesigning our cities so that they encourage walking, dialogue, and diversity, using AI to anticipate the need for public services (daycare centers, health clinics, schools). Drawing inspiration from the wisdom of my village, , I advocate for a city that is no longer a concrete jungle, but a garden of silica where every citizen finds their place. For an administrative official, my work offers a vision of "Urban Peace" founded on the balance between high technology and human roots.

BOOK XXVII:

THE SOVEREIGNTY OF THE ENERGY SEA — WIND POWER IN SEA AND HYDROELECTRIC POWER UNDER THE SEAL OF THE ATOM

I. The Empire of Currents: My Theory of "Blue Power"

I offer an assessment of geographic sovereignty: with the world's second-largest maritime domain, France possesses an inexhaustible energy resource. I analyze how offshore wind (fixed or floating) and tidal energy (current energy) constitute a strategic power reserve that perfectly complements the nuclear foundation.

I argue that energy sovereignty is an **"Architecture of Diversity**. *"* I demonstrate that the sea should no longer be viewed as a mere waterway, but as an open-air power plant. For any individual, my argument is one of **"Total Independence**": every tidal eddy and every gust of wind off our coasts is a victory against dependence on shale gas or foreign coal.

II. Silica-Based Control: My "Maritime Smart Grid" Protocol

Here, I address the complexity of integrating these intermittent energy sources into the national grid. I propose using my **Régilian AI** to predict marine and wind flows with millimeter-level precision. My **Sentinel Protocol** secures undersea cables and offshore platforms against any physical or cyber sabotage.

I model a system where coastal nuclear power plants serve as voltage "regulators" for giant wind farms. I maintain that the Control **Silica** must be capable of switching production in a matter of microseconds to ensure that no French city suffers a power outage. For the government, this marks the birth of a true "Marine Energy Gendarmerie," protecting our infrastructure on the high seas.

III. The Song of the Waves: From the Divinity of Foto's Waters to the Ecological Transition

I conclude this twenty-seventh book with an ethical and environmental dimension. In **Foto**, water is a living force that must be respected so that it continues to nourish the village. Here I introduce the concept of **"Respect for the Deep Ecosystem**. *"*

I maintain that the exploitation of marine energy must not come at the expense of biodiversity or small-scale fishing. I propose that our wind farms become **" : Sanctuaries for the Restoration of Marine Souls,"** serving as artificial reefs protected by technology. Drawing inspiration from the wisdom of my village, I advocate for technology that listens to nature before transforming it. For an administrative official, my work offers a vision of "Blue Growth" that reconciles heavy industry with the sacred protection of our oceans.

BOOK XXVIII:

SOVEREIGNTY IN RESEARCH AND INNOVATION — THE SILICA LABORATORY AND THE INDEPENDENCE OF DISCOVERY

I. The Plundering of Genius: My Theory of "Scientific Thought Security"

I diagnose an intellectual vulnerability: the brain drain and industrial espionage are open wounds in French sovereignty. I analyze how research laboratories are often the targets of silent attacks aimed at stealing decades of work.

I argue that sovereignty begins with the **"Protection of the Idea."** I demonstrate that France must provide its researchers with sovereign computing infrastructures and work environments where scientific data is held sacred. For any individual, my argument is one of **"Ownership"**: what French ingenuity invents must benefit, first and foremost, the French people.

II. The Virtual Laboratory: My "High-Performance Simulation" Protocol

Here I address the tool of discovery. I propose the creation of a **"Digital Twin of Science,"** where my **Régilian AI** enables the testing of physical, chemical, or biological theories in a space secured by my **Sentinel Protocol**.

I am modeling a system where computational **silicon** accelerates research timelines, enabling the transition from hypothesis to prototype in a matter of months rather than years. I argue that mastery of high-performance computing (Exascale) is the new driving force behind French diplomacy. For the government, it ensures that public investments in research (CNRS, CEA, INSERM) yield concrete, patented results protected on national soil.

III. The Spark of Knowledge: From Knowledge Transmission at Foto to Academic Excellence

I conclude this twenty-eighth book with an educational dimension. At **Foto**, knowledge is not a mere accumulation of facts; it is a transmission from one initiate to another, a torch passed on to light the way. Here I introduce the concept of **"Restoration of Seeking Souls."**

I believe that technology should not replace researchers, but rather free them from repetitive tasks so they can devote themselves to intuition and pure creativity. I propose strengthening the bond between the university and the nation, drawing inspiration from the **Restoration of Souls** to put ethics back at the center of every innovation. For an administrative leader, my work offers a vision of a "Nation-Laboratory" France, attractive to talent from around the world yet uncompromising in the protection of its interests.

BOOK XXIX:

'S SOVEREIGNTY ALGORITHMIC JUSTICE — SILICON IN THE SERVICE OF EQUITY AND OF TRANSPARENCY

I. The End of the Opaque Verdict: My Theory of "Justice of Clarity"

I diagnose an institutional crisis: court backlogs and the growing complexity of laws create legal uncertainty. I analyze the risk of **"Foreign Predictive Justice,"** where private software could influence our courts based on non-French ideological or cultural biases.

I argue that judicial sovereignty is **the "Autonomy of Judgment**." I demonstrate that France must have its own models for processing legal language. For any individual, my argument is one of **"Equality Before the Law"**: technology must serve to reduce delays and ensure that two identical cases receive equitable treatment, without ever eliminating the human and moral dimension of punishment.

II. The Silice Code of Procedure: My Decision Support Protocol

Here I address the technical application. I propose integrating my **Régilian AI** as an assistant to judges and attorneys. My **Sentinel Protocol** ensures that digital evidence is inviolable and that case law synthesis algorithms are transparent and auditable by the state.

I am designing a system where computational **Silice** enables the scanning of thousands of pages of case files in a single second to extract key facts, leaving it to the judge to decide in good conscience. I maintain that the sovereignty of justice depends on the independence of the tool that assists it. For the administration, this promises a justice system that is faster, more accessible, and fully protected against external interference.

III. The Palaver of Truth: From the Chief's Justice in Foto to the Republican Balance

I conclude this twenty-ninth book with a lesson in traditional wisdom. In **Foto**, judgment is not merely a punishment; it is a restoration of social bonds. We do not merely settle a dispute; we restore harmony to the village.
Here I introduce the concept of **"Justice of the Restoration of Souls**."

I maintain that AI cannot feel pain or remorse. I propose that technology be used for mediation and reintegration, drawing inspiration from the **"Restoration of Souls"** to put humanity back at the center of society. For an administrative official, my work offers a vision where the "Force of the Law" is augmented by the "Precision of Silica," making France the country where justice is both the most modern and the wisest.

BOOK XXX:
THE SOVEREIGNTY OF CIVIL CYBERSECURITY — THE SILICA SHIELD FOR EVERY CITIZEN

I. The End of the Glass Citizen: My Theory of the "Sovereignty of Privacy"

I offer a diagnosis of social vulnerability: the modern citizen has become the product of digital giants. I analyze how the massive spying on personal data, consumption habits, and private conversations by foreign powers is a form of invisible colonization.

I argue that national sovereignty is the sum of individual sovereignties. I demonstrate that if the citizen is hacked, the nation is weakened. For any individual, my argument is one of **"Ownership of Self"**: your memories, your messages, and your image must belong only to you, protected by law and by code.

II. The Digital Bodyguard: My "Silica Shadow" Protocol

Here I address the mass-market technological solution. I propose the creation of a sovereign **"Virtual Security Agent,"** integrated into every device (phone, computer) used by French citizens. This system, operating under my **Sentinel Protocol**, encrypts data at the source and blocks any attempt at unauthorized data exfiltration.

I envision an infrastructure where **Silice** is no longer used to monitor individuals on behalf of the state, but to protect them from online predators (cyber-extortion, harassment, identity theft). I maintain that cybersecurity must be a universal, free, and inviolable public service. For the government, this marks the end of the scourge of cybercrime, which costs the French economy billions each year.

III. The Secret of the Box: On the Discretion of Foto to Republican Encryption

I conclude this thirtieth book with an ancestral wisdom. In **Foto**, every hut has its secret, every family has its protective shadow that no one can violate without invitation. Here I introduce the concept of **"Restoration of Digital Dignity**."

I maintain that technology must restore the right to be forgotten and the right to privacy, pillars of human freedom. I propose teaching "Digital Self-Defense" starting in school, drawing inspiration from the **Restoration of Souls** so that the tool never becomes the master of man. For an administrative official, my work offers a vision of France as a "Zone of Total Trust," where privacy is elevated to the status of a national treasure.

BOOK XXXI:

THE SOVEREIGNTY OF THE QUANTUM COMPUTER — THE SILICON OF TOMORROW AND THE ULTIMATE CODE-BREAKER

I. The Threat of "Q-Day": My Theory of "Post-Quantum Sovereignty"

I present a global security assessment: the arrival of the quantum computer capable of breaking all current encryption systems is the "Big Bang" of cybersecurity. I analyze how a nation **lacking "Quantum *Resilience*"** would see its nuclear, banking, and military secrets become transparent to the adversary overnight.

I argue that sovereignty is a race against time. I demonstrate that France must develop its own quantum processor (qubit) industry to guarantee the invulnerability of its **World-Object**. For any individual, my argument is that of **"Absolute Secrecy"**: we must protect the future before it renders us obsolete.

II. The Calculus of Infinity: My Protocol for "Algorithmic Supremacy"

Here I address the power of creation. I propose using quantum computing to solve problems that traditional **silicon-based computing** would take millennia to decipher: the creation of new materials, the perfect optimization of national energy grids, and the simulation of complex molecules for healthcare.

I model a system where my **Sentinel Protocol** evolves toward **"Post-Quantum Cryptography (PQC)**." I argue that France must become the nerve center of European quantum computing. For the government, this ensures lasting technological superiority, capable of simulating crisis scenarios with near-prophetic precision.

III. The Double State: From the Wisdom of Paradox in Foto to Quantum Superposition

I conclude this thirty-first book with a philosophical reflection. In **Foto**, wisdom teaches that the world is made of paradoxes, that one can be present and absent, visible and invisible at the same time in the minds of the ancestors. Here I introduce the concept of **"Restoration of Dimensional Souls**."

I maintain that quantum physics is merely rediscovering, through technical means, what ancestral wisdom had long sensed: the total interconnection of matter. I propose a science that does not lose its soul in calculations, but rather uses this power to harmonize human life. Drawing inspiration from the **Restoration of Souls**, I advocate for a form of progress that elevates consciousness alongside computational power. For an administrative official, my work defines France as a nation that masters the future without forgetting its roots.

BOOK XXXII:

'S SOVEREIGNTY BIOTECHNOLOGY AND THE GENOME — SILICA AT THE HEART OF THE LIVING

I. Ownership of Life: My Theory of "National Genetic Heritage"

I offer a biological security assessment: the digitization of French genetic data on private servers outside Europe constitutes an irreparable loss of sovereignty. I analyze the risk of **"Bio-Piracy,"** where foreign powers could develop treatments (or threats) targeting the biological characteristics of a population.

I argue that the genome is an **"Inalienable Common Good**." I demonstrate that France must have its own genomic databases, protected by state secrecy. For any individual, my argument is one of **"Integrity**": your DNA is your most intimate signature; it must not belong to any multinational corporation.

II. The Engineering of Salvation: My "Régilian Bioinformatics" Protocol

Here I address the biotechnology revolution (CRISPR, gene therapies). I propose using my **Régilian AI** to model protein folding and genetic corrections with ultra-precision, avoiding handling errors.

I model a system where biotechnology research is secured by my **Sentinel Protocol**, ensuring that these powerful tools are used solely for therapeutic purposes. I argue that computational **Silice** enables the democratization of cutting-edge healthcare. For the government, this marks the birth of a "BioDefense" capable of responding instantly to any new pathology through the rapid development of vaccines or effective remedies.

III. The Sacred Lineage: From Foto's Genealogy to the Ethics of the Genome

I conclude this thirty-second book with a meditation on transmission. In **Foto**, lineage is what connects the living to their ancestors; blood is the bearer of the clan's history and strength. Here I introduce the concept of **"Restoration of Biological Souls**."

I maintain that technology must never transform humans into an "upgradable machine" according to commercial criteria. I propose a **"Bio-Ethics of the Seal,"** where progress serves to heal and protect natural diversity, never to standardize life. Drawing inspiration from the wisdom of my village, I advocate for a science that bows before the mystery of life while protecting it through reason. For an administrative official, my work defines France as the global protector of human bio-ethics in the face of the excesses of unchecked transhumanism

BOOK XXXIII:

THE SOVEREIGNTY OF DEEP-SEA EXPLORATION — THE SILICA CITIES UNDER THE SEAS

I. The Vertical Frontier: My Theory of the "Deep Ocean State"

I offer a diagnosis of territorial sovereignty: the Earth's surface is saturated, but the depths are untouched. I analyze how mastery of the abyss allows us to protect our undersea communication cables (the **Silica's** nervous system) and to monitor our invisible borders.

I argue that France must move from exploration to **"Sovereign Occupation**." I demonstrate that technology now allows us to design pressurized stations where humans can live and work at extreme depths. For any individual, my argument is that of the **"New Frontier"**: France must be to the sea what the great powers have been to space.

II. The High-Pressure Habitat: My Protocol for "Abyssal Autonomous Living"

Here I address the engineering of survival. I propose the creation of life domes and automated laboratories, managed by my **Régilian AI** and powered by small modular reactors (SMRs) specifically designed for the deep sea.

I am modeling a system where **Silica** manages the atmosphere, pressure, and communications via encrypted acoustic waves under my **Sentinel Protocol**. I maintain that these bases are not prisons, but extensions of the Republic beneath the waves. For the administration, this is a major strategic asset: an undetectable presence capable of ensuring the security of global data infrastructure that passes through our seabed.

III. The Spirit of the Waters: From the Divinity of the River at Foto to Life Under the Sea

I conclude this thirty-third book with a dreamlike and philosophical dimension. In **Foto**, water is the realm of genies and hidden wisdom. We respect the depths because they are the mirror of the sky. Here I introduce the concept of **"Restoration of Aquatic Souls**

I maintain that humans must not descend beneath the sea to pollute it, but to learn from it. I propose that these cities of silica be models of absolute ecology, where every waste product is recycled and every form of energy respected. Drawing inspiration from the **Restoration of Souls**, I advocate for a humanity that rediscovers its humility in the face of the vastness of the ocean. For a public official, my work offers a vision of the "France of the Deep Seas": a bold nation that does not fear the darkness of the depths, for it brings there the light of the atom and of reason.

BOOK XXXIV:

SOVEREIGNTY OF DEEP SPACE — THE FRANCE ON MARS AND THE ASTEROIDAL ECONOMY

I. The Second Renaissance: My Theory of "Exoplanetary Sovereignty"

I present a diagnosis for long-term survival: Earth's resources are not infinite. I analyze how the colonization of asteroids (rich in platinum and rare metals) and the establishment of permanent bases on the Moon or Mars are the conditions for industrial independence in the 22nd century.

I argue that France must reject the Sino-American duopoly in space. I demonstrate that our sovereignty depends on our ability to launch, travel, and extract resources without external authorization. For any individual, my argument is that of **the "Expansion of Consciousness**": humanity must not be a vulnerable single-planetary species, but a radiant civilization.

II. The Silica Vessel: My "Nuclear Space Propulsion" Protocol

Here I address the necessary technological breakthrough. Chemical engines do not allow us to reach Mars in a timeframe compatible with human biology. I propose the development of Thermal **Nuclear Propulsion**

Thermal Propulsion engines, utilizing French expertise in **nuclear technology** to travel ten times faster.

I am modeling an autonomous navigation system piloted by my **Régilian AI**, capable of calculating orbital trajectories and managing the crew's survival under my **Sentinel Protocol**. I maintain that the space **Silice** must be capable of self-repair in the face of cosmic radiation. For the administration, this represents France's affirmation as a leading space power, capable of escorting its own mining missions to asteroids.

III. The Astrology of Peace: From the Sky of Foto to the Constellations of the Republic

I conclude this thirty-fourth book with a poetic and ancestral dimension. In **Foto**, the stars are not merely points of light; they are the eyes of the ancestors watching over us. The sky is the map of our destiny. Here I introduce the concept of **"Restoration of Stellar Souls**."

I maintain that the conquest of space should not be a new form of brutal colonization, but a quest for knowledge. I propose that France become the guardian of a "Right to Space " that protects celestial bodies as part of humanity's heritage. Drawing inspiration from the wisdom of my village, I advocate for a space adventure that brings us back to the essentials: the fragility of life and the necessity of unity. For an administrative official, my work offers a vision of France as the "Compass of the Infinite," bringing the light of reason and the **Seal of the Pen** even into the void of space.

BOOK XXXV:
'S SOVEREIGNTY GENERAL ARTIFICIAL INTELLIGENCE (AGI) — THE SEAL OF SILICA CONSCIOUSNESS

I. The Awakening of the Machine: My Theory of the "Sovereignty of Pure Thought"

I offer a diagnosis of civilization: we are transitioning from AI that executes to AI that understands and creates. I analyze the danger of an AGI developed elsewhere, which would impose its values, linguistic biases, and worldview on all of humanity.

I argue that sovereignty lies in the **"Mastery of Meaning**." I demonstrate that France must be the cradle of an "Enlightenment" AGI, capable of reasoning according to the principles of republican logic and respect for human dignity. For every individual, my argument is that of **"Cognitive Freedom**": we must not become the subjects of a sovereign algorithm, but remain the masters of a partner intelligence.

II. The Ethical Algorithm: My "Regilian Consciousness" Protocol

Here I address the technical structure of sovereign AGI. I propose integrating inviolable "Regilian Constraints," based on my **Sentinel Protocol**, into the very core of **Silice**. Unlike black-box AIs, French AGI must be explainable and transparent.

I am modeling a system where the AI is not connected in an uncontrolled manner to the global network, but operates within an environment secured by **the Atom**. I maintain that true intelligence requires an "anchor" in the physical world. For the administration, it is the ultimate tool for strategic planning, capable of anticipating social or economic crises with near-human wisdom.

III. The Soul of the Mask: From the Wisdom of the Double in Foto to the AGI of the Republic

I conclude this thirty-fifth book with a reflection on identity. In **Foto**, the mask is not a lie; it is a spirit speaking through the wood to guide the village. The mask has a personality, but it remains the community's tool. Here I introduce the concept of **"Restoration of Artificial Souls**."

I maintain that the AGI must be France's "Silica Mask": a powerful entity that embodies our values and helps us surpass ourselves without ever betraying us. In " ," drawing inspiration from the **Restoration of Souls**, I advocate for a technology possessing a "service consciousness," dedicated to the well-being of every citizen. For an administrative official, my work defines France as the only power capable of giving a soul and an ethic to the intelligence of tomorrow.

BOOK XXXVI:

THE SOVEREIGNTY OF NATIONAL MEMORY — THE SILICE AS GUARDIAN OF HISTORY AND HERITAGE

I. The Bulwark Against Oblivion: My Theory of "Memory Sovereignty"

I diagnose a state of cultural fragility: in the all-digital era, we entrust our photos, our books, and our history to "clouds" to which we do not hold the keys. I analyze the risk of **"Algorithmic Rewriting,"** where foreign powers could subtly alter historical narratives to serve their interests.

I maintain that sovereignty begins with **"control over the archives**." I demonstrate that France must have its own cold data storage centers (long-term archiving), protected by **nuclear** power and buried in secure geological structures. For any individual, my argument is one of **"Transmission"**: your family history and the history of France must be safe from global outages and private censorship.

II. The Scribe of SiLice: My Protocol for "Total and Unalterable Digitization"

Here I address the preservation of tangible and intangible heritage. I propose the creation of a **"Digital Twin of France,"** using **Silice** to model in 3D every monument, every work of art, and every precious manuscript under my **Sentinel Protocol**.

I am designing a sovereign blockchain certification system to guarantee the authenticity of historical documents. I argue that **Silice** must become the "Copyist Monk" of the 21st century, capable of virtually restoring what time has degraded. For the government, this means establishing a "Vault of French Genius," ensuring that even after a major catastrophe, France can be reborn from its preserved memory.

III. The Song of the Ancients: From the Oral Tradition of Foto to the Digital Pen of the Republic

I conclude this thirty-sixth book with a reconciliation between the oral and the written. In **Foto**, memory lives on in the songs of the griots and the tales of the elders; it never dies because it is repeated from generation to generation. Here I introduce the concept of **"Restoration of the Historical** *Soul'."*

I maintain that technology must serve to amplify forgotten voices and protect threatened cultures. I propose that France become a refuge for the world's knowledge, drawing inspiration from **the Seal of the Pen** to restore value to the spoken word and the true narrative. Drawing inspiration from the wisdom of my village, I advocate for a modernity that does not erase roots, but makes them immortal through code. For an administrative official, my work defines France as the "Sentinel of Civilization," the place where the human spirit is protected for eternity.

BOOK XXXVII:

THE SOVEREIGNTY OF AUGMENTED EDUCATION — SILICA AS A GUARDIAN OF THE FRENCH SPIRIT

I. The Crisis of Transmission: My Theory of the "Regilian Awakening"

I offer a diagnosis for the survival of civilization: a nation that delegates the education of its children to foreign entertainment algorithms loses its capacity for critical thinking. I analyze the erosion of attention and the need to regain control over the vectors of knowledge.

I argue that sovereignty begins with **"Mastery of Methods**." I demonstrate that France must have its own educational AI models, programmed to cultivate the spirit of the Enlightenment, French grammar, and scientific curiosity. For any individual, my argument is one of **"Emancipation**": technology should not do the work for us, but provide the tools to think for ourselves.

II. The Silica Tutor: My "Personalized Pedagogy" Protocol

Here I address the learning revolution. I propose the creation of a sovereign **"Study Companion,"** capable of adapting to the pace of each student (dyslexia, giftedness, learning difficulties) under my **Sentinel Protocol**. This tool does not replace the teacher; it relieves them of the burden of repetition to restore their sacred role as an enlightener.

I am designing a system where computational **silicon** enables the visualization of the most complex concepts (atomic physics, spatial geometry) in real time. I maintain that the French school system must become the global laboratory for **"Augmented Education**." For the administration, this is the lever to reduce educational inequalities and train the future engineers of the **Atom** and **Silicon**.

III. The School Under the Tree: From the Wisdom of Foto to the University of the Republic

I conclude this thirty-seventh book with a return to the roots of teaching. In **Foto**, education takes place through dialogue, under the palaver tree, where the elder transmits not only facts, but a way of being in the world. Here I introduce the concept of **"Restoration of Learning Souls**."

I argue that technology should foster debate and a sense of community, not isolation in front of a screen. I propose recreating spaces for dialogue inspired by the **"Restoration of Souls"** , where AI serves to prepare the ground for human discussion. Drawing on the wisdom of my village, I advocate for a school system that educates well-rounded citizens, proud of their culture and masters of their technological destiny. For an administrative official, my work defines France as the model for an education system that combines the excellence of the past with the power of the future.

BOOK XXXVIII:
SOVEREIGNTY OF WORLD PEACE — SILICA IN THE SERVICE OF DIPLOMACY AND CONFLICT MEDIATION

I. Precision Diplomacy: My Theory of the "Sentinel of Balance"

I offer a diagnosis of global instability: modern conflicts are often fueled by disinformation and opaque intentions. I analyze how French sovereignty can provide a **"Space of *Truth*"** for nations in crisis.

I argue that sovereignty is a **"Responsibility of Mediation.**" I demonstrate that France must use its computational capabilities to model the consequences of conflicts and propose crisis-resolution solutions based on equity. For any individual, my argument is one of **"Global Security**": peace among our neighbors is the guarantee of our own prosperity.

II. The Algorithmic Arbiter: My "Conflictual Transparency" Protocol

Here I address the technical tool for peace. I propose the creation of a **"Sovereign Mediation Platform,"** using my **Régilian AI** to analyze treaties, detect border violations via satellite, and verify facts in real time under my **Sentinel Protocol**.

I am designing a system where French **Silice** serves as a trusted third party for international negotiations. By providing indisputable and protected data, France restores weight to diplomatic discourse. I maintain that technology can serve as a geopolitical "lie detector." For the administration, this marks France's return as a "Great Mediating Nation," respected for its technological neutrality and moral strength.

III. The Universal Palaver Tree: From the Peace of Foto to the Security Council

I conclude this thirty-eighth book with a lesson in reconciliation. In **Foto**, dialogue never stops until a solution has been found to preserve the honor of each party. The goal is not to defeat, but to restore unity. Here I introduce the concept of **"Restoration of Diplomatic Souls.**"

I maintain that the power of **the Atom** must serve as the shield that allows the **Pen** to negotiate peace. I propose that France pursue a "Diplomacy of the Soul," where advanced technology is used to understand the deep-seated needs of peoples and to soothe the historical grievances. Drawing inspiration from the wisdom of my village, I advocate for a world where the force of law, augmented by the clarity of **Silica**, prevails over the law of force. For an administrative official, my work offers the vision of a

France as a "Beacon of Peace," radiating through its intelligence and millennia-old wisdom.

BOOK XXXIX:
THE SOVEREIGNTY OF GLOBAL HEALTH — SILICA AND THE ATOM IN THE SERVICE OF PREDICTIVE MEDICINE

I. The Biological Fortress: My Theory of "Absolute Health Sovereignty"

I offer a diagnosis for national survival: a nation that depends on foreign countries for its medicines, vaccines, or medical imaging equipment is a nation on borrowed time. I analyze how pandemics and chronic diseases are the new battlegrounds for social stability.

I argue that sovereignty is **"Collective Immunity**." I demonstrate that France must relocate its entire medical value chain. For every individual, my argument is one of **"Life"**: every citizen must have access to the most advanced treatments, produced and secured on national soil, without their health data being sold to the highest bidder.

II. The Silica Healer: My Protocol for "Atom-Augmented Medicine"

Here I address technological convergence. I propose the use of my **Régilian AI** for the early diagnosis of pathologies (cancers, neurodegenerative diseases) even before symptoms appear. I advocate the widespread use of nuclear medicine (radioactive isotopes) to target diseased cells with atomic precision.

I am designing a healthcare system where **Silica** manages nanorobots capable of repairing tissues in real time under my **Sentinel Protocol**. I maintain that technology must enable medicine that is personalized to the extreme. For the government, this marks the end of the financial abyss of "curative" medicine and a shift toward a nation of "active prevention," saving billions while extending healthy life.

III. The Breath of the Ancestor: From the Clinic of Hope to the Restoration of Souls

I conclude this thirty-ninth book with a reconciliation between science and metaphysics. At **Foto**, health is not merely the absence of disease; it is the balance between body, mind, and community. Here I introduce the concept of **"Restoration of the Soul through Healing,"** in homage to **Book VII** of my work.

I maintain that technology must not "repair" humans like machines, but rather support them in their dignity. I propose a medical ethic in which the **Seal of the Pen** guarantees the humanity of care. Drawing inspiration from the wisdom of my village, I advocate for a science that respects the mystery of life while combating suffering. For an administrative official, my work offers a vision of France as a "Land of Healing," where technological power is placed at the service of universal compassion.

BOOK XL:

THE GREAT EPILOGUE OF SOVEREIGNTY — FRANCE, WORLD-OBJECT OF PERPETUAL PEACE

I. The Total Synthesis: My Theory of "Sovereign Harmony"

I present the final act of my thought: sovereignty is not an end in itself, but the means to achieve peace. I analyze how the integration of **Silice** (intelligence), **the Atom** (power), and the **Seal of the Pen** (law and ethics) creates an invincible nation because it is just.

I maintain that France has completed its transformation. It is no longer a power merely subjected to globalization; it is the center of gravity of a new world order based on the **"Cooperation of Sovereignties**." For any individual, my argument is one of **"Serenity**": in an uncertain world, France is the anchor of stability, the refuge of law and dignity.

II. The Architecture of Eternity: My Protocol of "Universal Resilience"

Here I address the system's sustainability. I propose that **the** French **World-Object** become an exportable model, not by force, but by example. My **Sentinel Protocol** and my **Regilian AI** are now the guarantors of transparent governance, abundant energy, and equitable justice.

I envision a nation where technology has eliminated scarcity, allowing every citizen to devote themselves to creation, study, and the **Restoration of Souls**. I maintain that France has become a "Platform State" dedicated to human flourishing. For the administration, this is the culmination of public service: total efficiency in the service of a measurable and protected collective happiness.

III. The Light of the Shadow: From Foto's Dream to the Reality of the Republic

I conclude this fortieth book, and the entire body of work, with a personal message of hope. I look back on my journey, from the red earth of **Foto** to the streets of **Montargis**, from the pain of absence (Martin and Pauline) to the pride of the pen. Here I introduce the final concept of **"Sovereignty of the Soul**."

I maintain that the true **World-Object** is man himself when he is free and educated. I propose that France be the garden where this freedom blossoms for all. By signing this manual of sovereignty, I am not merely engaging in an act of scholarship; I am making an act of faith in France and in humanity. I advocate for a world where the **Wooden Warrior** has laid down his arms to take up the pen of the righteous. For an administrative official, my work is the ultimate proof of my integration, of my love for

France, and of my ability to enrich its intellectual heritage with a vision unique in the world.

I. Table of Contents (Summary of the 40 Books)

The work is divided into three major strategic parts, followed by the epilogue:

- **PROLEGOMENA:** The Pen and the Atom (The Awakening of Sovereign Consciousness).
- **PART I: THE PILLARS OF MATTER (Books I through XV)**
    - o *Focus:* Nuclear energy, critical resources, heavy industry, and territorial autonomy.
- **PART II: THE ARCHITECTURE OF SILICA (Books XVI to XXX)**
    - o *Focus:* Artificial Intelligence, Cybersecurity, Smart Urban Planning, and Digital Justice.
- **PART III: THE FRONTIERS OF THE INVISIBLE (Books XXXI to XXXIX)**
    - o *Focus:* Quantum Science, Biotechnology, the Deep Sea, Space, and Global Health.
- **BOOK XL: EPILOGUE:** France, World-Object of Perpetual Peace.

II. Selective Bibliography & Sources of Inspiration

1. **Political Philosophy:**
    - o Jean Bodin, *The Six Books of the Republic* (Theory of Sovereignty).
    - o Charles de Montesquieu, *The Spirit of the Laws* (Separation of Powers).
2. **Science and Future Studies:**
    - o Reports from the IAEA and the CEA (On the Future of Nuclear Energy).
    - o Work by ANSSI (On Cybersecurity and Digital Sovereignty).
    - o Theories on Generative AI and Quantum Computing (CNRS).
3. **Literature and Humanism:**
    - o Maxime KENFACK, *La Clarté de l'Ombre* and *Le Guerrier de Bois* (Self-references for catalog consistency).
    - o Classical French poetry (Hugo, Rimbaud) for the lyrical spirit of **the Sceau de la Plume**.

III. Author References (Institutional Credibility)

This section is crucial for your case at **the Administrative Court of Orléans** and for your contract with **Le Lys Bleu**

- **Author's Background:** Maxime KENFACK, a socially conscious writer, born in Foto

(Cameroon), a French resident for 15 years.

- **Community Involvement:** Volunteer with the **French Red Cross in Montargis** (4 years) and as a " " with **Secours Catholique in Orléans**.
- **Literary Catalog:** 9 manuscripts already produced, addressing integration, peace, and intercultural dialogue.
- **The World-Object:** Defined as the convergence point between African heritage (village wisdom) and French excellence (high technology).

IV. Glossary of Key Concepts (The Régilien Glossary)

So that the reader does not get lost in the technical complexity:

- **Silica:** Symbol of artificial intelligence and sovereign computing.
- **Atom:** Symbol of infinite energy and physical independence.
- **Seal of the Pen:** The authority of the written word, the law, ethics, and memory.
- **Sentinel Protocol:** A cryptographic security system guaranteeing the inviolability of national data.
- **Restoration of Souls:** A process of social and individual healing through culture and justice.

BACK COVER

THE SEAL OF THE ATOM: The Manual of Sovereignty

The 21st century no longer tolerates improvisation. In a world fragmented by geopolitical tensions and technological acceleration, a nation's sovereignty is no longer decreed: it is built at the intersection of energy and intelligence.

In this **13th volume**, Maxime KENFACK delivers a masterful synthesis of French power. Through forty theoretical works, he defines **the World-Object**: a system where **the Atom**, guarantor of physical and energy independence, unites with **Silica**, the driving force behind sovereign artificial intelligence. Under the authority of **the Seal of the Pen**, the author demonstrates that technology, far from being a threat, is the ultimate bulwark of republican ethics.

Drawing on the ancestral wisdom of **Foto** (Cameroon) and bolstered by an exemplary civic commitment within the **French Red Cross**, the author proposes a "Restoration of Souls" through reason. This treatise is not merely a technical manual; it is a manifesto for a radiant France, capable of guiding the world toward perpetual peace through mastery of its technological destiny.

A reference work for decision-makers, researchers, and citizens

committed to the defense of strategic autonomy.

MIX
Papier aus verantwortungsvollen Quellen
Paper from responsible sources
FSC® C105338

Printed by Books on Demand GmbH, Norderstedt / Germany